Ceramic Materials - Synthesis, Characterization, Applications and Recycling

Edited by Dolores Eliche Quesada, Luis Perez Villarejo and Pedro Sánchez Soto

Published in London, United Kingdom

IntechOpen

Supporting open minds since 2005

Ceramic Materials - Synthesis, Characterization, Applications and Recycling
http://dx.doi.org/10.5772/intechopen.75375
Edited by Dolores Eliche Quesada, Luis Perez Villarejo and Pedro Sánchez Soto

Contributors
Shukdev Pandey, Devendra Kumar, Om Parkash, Lakshman Pandey, Jin Sam Choi, Suminar Pratapa, Ella Kiswanti, Dien R Diana, Yufi Hariyani, Lisma Dk Sari, Musyarofah Musyarofah, Tri Wikantoro, Malik A Baqiya, Amr El-Dieb, Mahmoud Taha, Samir Ibrahim Abu-Eishah, Dolores Eliche Quesada

First published in London, United Kingdom, 2019 by IntechOpen
IntechOpen is the global imprint of INTECHOPEN LIMITED, registered in England and Wales, registration number: 11086078, The Shard, 25th floor, 32 London Bridge Street
London, SE19SG - United Kingdom
Printed in Croatia

British Library Cataloguing-in-Publication Data
A catalogue record for this book is available from the British Library

Additional hard copies can be obtained from orders@intechopen.com

Ceramic Materials - Synthesis, Characterization, Applications and Recycling
Edited by Dolores Eliche Quesada, Luis Perez Villarejo and Pedro Sánchez Soto
p. cm.
Print ISBN 978-1-78985-773-3
Online ISBN 978-1-78985-774-0

Meet the editors

Prof. Dra. Dolores Eliche Quesada is an associate professor of the Materials Science and Metallurgical Engineering Area of the University of Jaén (Spain), a chemical engineer in the University of Granada (Spain), and has a PhD in Sciences from the University of Malaga (Spain). Her scientific activity has focused on materials science and technology, and in particular on the development of new sustainable materials through the valorization of industrial waste and biomass for the construction sector, the characterization of raw materials, the determination of the properties of ceramic materials, and the innovation of products. She is the author of 45 scientific publications indexed mostly in the first quartile of the ISI, and 10 book chapters, as well as more than 100 contributions to national and international scientific conferences.

Prof. Dr. Luis Pérez Villarejo is an associate professor of the Materials Science and Metallurgical Engineering Area of the University of Jaén (Spain), a chemical engineer at the University of Jaén (Spain), a materials engineer at the University of Almería. He has an MSc in Science and Technology of New Materials from the University of Seville and a PhD in Chemistry from the University of Jaén. His research is focused on the study of ceramic materials from industrial by-products and the study of their properties, geopolymeric and cementitious systems, ceramic foams, as well as nanomaterials. He is the author of 32 scientific publications indexed mostly in the first quartile of the ISI, and 14 book chapters, as well as almost 100 contributions to national and international scientific conferences.

Dr. Pedro J. Sánchez-Soto is a research scientist at CSIC. He has an MSci (1982), Degree (1985), and PhD in Chemistry (1990) from the University of Sevilla (Spain). He is also a postdoctoral researcher (1990–1991) at the University of Florida (USA). He joined the ICMS in 1992 as a tenured scientist of CSIC, being appointed a research scientist in 2007. He has conducted several research projects and published more than 200 SCI articles, book chapters, books, and other non-SCI articles. He has supervised six PhD theses. He is the coeditor or author of more than 10 books and 10 patents. His main research interests are the processing of ceramics, raw materials and vitreous ceramics, including clays and cultural heritage, and the relationships between their microstructure and properties. He is a member of several scientific societies (American Ceramic Society) and an associate editor of *Boletín de la Sociedad Española de Cerámica y Vidrio.*

Contents

Preface **XI**

Chapter 1 **1**
Introductory Chapter: Ceramic Materials - Synthesis, Characterization, Applications and Recycling
by Dolores Eliche-Quesada,Luis Pérez-Villarejo and Pedro José Sánchez-Soto

Chapter 2 **7**
The Use of Ceramic Waste Powder (CWP) in Making Eco-Friendly Concretes
by Amr S. El-Dieb, Mahmoud R. Taha and Samir I. Abu-Eishah

Chapter 3 **43**
Synthesis of High-Purity Ceramic Nano-Powders Using Dissolution Method
by Suminar Pratapa, Ella A.D. Kiswanti, Dien R. Diana, Yufi Hariyani, Lisma D.K. Sari, Musyarofah Musyarofah, Triwikantoro Triwikantoro and Malik A. Baqiya

Chapter 4 **59**
Plasma Resistance Evaluation and Characteristics of Yttria Ceramics Sintered by Using Calcination Yttria
by Jin Sam Choi

Chapter 5 **79**
Impedance Spectroscopy: A Powerful Technique for Study of Electronic Ceramics
by Shukdev Pandey, Devendra Kumar, Om Parkash and Lakshman Pandey

Preface

Scientific and technological development has led to the formulation of tailor-made materials, which have given rise to materials with new structural and industrial applications. This book aims to analyze the synthesis, characterization, and applications of ceramic materials. First, an introduction to traditional and advanced ceramics is made, followed by an overview of material preparation techniques and various applications in recent years. Then, traditional ceramic materials as ideal candidates for absorbing wastes are studied, offering the possibility of obtaining new eco-efficient ceramic materials, such as concrete, while competitively approaching the concept of the circular economy. The following chapters focus on the study of different advanced ceramics. During the last decade, nanoceramics have received significant attention as candidate materials due to their ability to demonstrate improved and unique properties compared to conventional ceramics. The obtaining of ceramic nanopowders is studied by a new obtaining method called the dissolution method with advantages with respect to solvothermal, sol-gel, and coprecipitation methods. Also analyzed are plasma resistance and the characteristics of yttria ceramics in terms of calcination and three-step sintering compared to alumina and zirconia ceramics as well as the characteristics of hot-pressed pure yttria ceramics that are annealed in an oxidation atmosphere. Finally, a study of some equivalent circuit models most useful for impedance spectroscopic studies of electronic ceramics and their simulated immittance behavior is made.

Dolores Eliche-Quesada
Department of Chemical, Environmental, and Materials Engineering,
Higher Polytechnic School of Jaén, University of Jaen,
Campus Las Lagunillas s/n, 23071 Jaén, Spain

Luis Pérez-Villarejo
Department of Chemical, Environmental, and Materials Engineering,
Higher Polytechnic School of Linares, University of Jaen,
Campus Científico-Tecnológico Cinturón Sur s/n, 23700 Jaén, Linares, Spain

Pedro J. Sánchez-Soto
Materials Science Institute of Sevilla (ICMS),
Joint Center Spanish National Research Council (CSIC)-University of Sevilla,
c/Américo Vespucio, 49, 41092 Sevilla, Spain

Chapter 1

Introductory Chapter: Ceramic Materials - Synthesis, Characterization, Applications and Recycling

Dolores Eliche-Quesada, Luis Pérez-Villarejo and Pedro José Sánchez-Soto

1. Introduction

Ceramic materials can be defined as inorganic materials constituted by the combination of metallic and nonmetallic elements whose properties depend on the way in which these elements are linked [1, 2]. Ceramic materials are the most versatile branch of materials. The origin of this versatility lies in the chemical nature of its bonds, since they are mainly constituted by strong ionic and covalent bonds in different proportions. The bonds determine a series of particular properties of ceramic materials among which are relatively high fusion temperatures, high modulus, high wear strength, poor thermal properties, high hardness and fragilities combined with tenacities, and low ductility. In addition to the lack of conduction electrons since they are combined forming chemical bonds, they are good electrical insulators.

Ceramic materials can be divided into two large groups: traditional ceramics and technical or advanced ceramics. Traditional ceramics can be defined as those that are based on silicates, among which are cement, clay products, and refractories. Traditional ceramics are produced in large volumes and constitute an important market. Traditional ceramic materials are made with raw materials from natural deposits such as clay materials. The second group, technical or advanced ceramics, is manufactured with artificial raw materials that have undergone an important chemical processing to achieve a high purity and an improvement of their physical characteristics. Therefore, they are manufactured with more advanced and sophisticated methods. Among them are carbides, nitrides, borides, pure oxides, and a great variety of ceramics with magnetic, ferroelectric, piezoelectric, and superconducting applications, among others. These ceramics possess excellent mechanical properties under extreme conditions of tension, high wear strength or excellent electrical, magnetic, or optical properties, or exceptional strength to high temperatures and corrosive environments, showing high strength to chemical attack [3]. There is a third group that is glasses that, although considered ceramic, are studied separately because they differ from the first group in the order reached by their crystalline structures as glass-ceramics.

The versatility mentioned above also allows ceramics to be used for a large number of end user and applications for the construction and building industry such as clay bricks and blocks, sanitary ware, and wall and floor tiles; in household goods like tableware (porcelain), cooking ware (glass-ceramics), or pottery; in industrial processing and manufacturing industry for refractories, filters, or cutting tools; or high technology applications as components for solid oxide fuel cells, nuclear ceramics as combustion elements (UO_2), coating reactors (AlN-SiC), and ceramic batteries. Due to their electrical properties, they can be useful as capacitors ($BaTiO_3$), high transition temperature superconductors ($YBa_2Cu_2O_7$), piezoelectric and ferroelectric, insulators, varistors (ZnO), and integrated circuit substrates. Due to their optical properties, they are used as infrared windows and construction of lasers and high-pressure sodium lamps. For its magnetic properties, the ceramic materials belonging to the family called ferrites stand out. They are usually grouped into two large groups: soft ferrites (spinels and garnets) and hard ferrites (hexaferrites), such as barium hexaferrite, spinel ferrites of Mn-Zn, or Ni-Zn ferrites doped with ruthenium or lanthanides that have a high saturation magnetization value and that have applications for permanent magnets, colored pigments, and electronics (such as interference suppressors, electronic amplifiers, or power inductors) or applications in microwave devices. Due to their hardness, they find utility in applications where abrasion properties such as those exhibited by silicon carbide or tungsten carbide (WC) are required, as well as for cutting tool applications. They can also be used in the defense field as armors or ballistic. On the other hand, many of them are biocompatible and can be used in the health field as 20 ceramic prostheses, dental implants, coatings for tissue ingrowth, access and fixation devices, etc. [4, 5].

One of the most important limitations of ceramic materials in structural applications is the fragility, due to the fact they tend to fracture in a fragile way with hardly any absorption of energy, besides they lack capacity for plastic deformation, especially under traction. This fragility to the fracture can be explained to a great extent due to the presence (something almost impossible to eliminate totally in this type of materials) of defects in the material that intensifies the same, being the ceramic materials much more resistant to compressive forces than to tension forces. This behavior is a consequence of the difficulty of movement of the dislocations by the ceramic structures, even at high temperatures. However, one of the uses that is currently increasing is the structural components that improve the mechanical properties. These uses require materials with high strength in various environmental conditions, capable of withstanding high temperatures and resistant to corrosion and oxidation. In addition, ceramic materials offer a substantial reduction in weight compared to other materials such as metals.

Manufacturing of the ceramic components involves the synthesis of the powder, mixing, shaping, and thermal treatment in which the sintering takes place either by diffusion in solid state or by formation of intergranular-liquid phase [6]. Ceramic powders are obtained either from the treatment of natural raw materials (rock minerals) or through synthetic routes such as sol-gel, self-propagated synthesis at high temperature (SHS), precipitation, etc. Generally the ceramic powders obtained from minerals are going to be used in traditional ceramics, while the particulate systems obtained by synthesis routes will have greater application in advanced ceramics, where the control of impurities and defects is more important in the final application of the piece, since it requires an exhaustive control of the purity of the raw materials and their final microstructure. The starting materials for traditional ceramics need previous processes for the elimination of impurities and the reproducible method to obtain a powder that can be shaped and sintered to obtain a piece. On the other hand, advanced ceramics use high-purity products as starting powders, obtained mostly by synthetic routes. The routes of obtaining ceramic materials are very varied, but the synthetically produced powders have common characteristics, such as high purity, uniform particle size distribution, high specific surface area, absence of hard agglomerates or aggregates that cannot be broken by grinding, and greater uniformity in the characteristics.

The shaping is the stage of processing by which a ceramic powder, duly prepared by washing processes, homogenization, etc. is compacted and conferred into a desired shape with high porosity and poor mechanical properties. The main existing forming processes are axial and isostatic dry pressing, porous mold or tape molding, or extrusion or injection molding. Each of these forming processes are used depending on the type of piece to be obtained and the time and costs in the production systems. The objective of the forming processes is generally to achieve high particle packing and low porosity (which will be partially or totally eliminated during the sintering process). The packaging, in addition to having a high number of coordination, should be as homogeneous as possible to avoid microstructural defects, which will mean a decrease in the properties of the final material. To do this, a series of parameters must be controlled, such as the particle size and its distribution, the level of agglomeration, and the flow of the powder during the forming process. There are other nontraditional methods for obtaining green parts that do not start from the ceramic powder or that after forming require some type of treatment before sintering, such as the sol–gel method, hot pressed (axial and isostatic), SHS, deposition of molten particles, and deposition in vapor phase (CVD) [5].

Once the green piece is obtained with the desired shape, before the densification of the material, it is necessary to proceed with the drying and the elimination of organics. In the drying step, the solvent used is removed by means of transport and evaporation in an unsaturated gaseous medium. In wet-formed materials, the solvent evaporates after shaping. If left unchecked it can generate defects and break the piece due to two main causes, namely, rapid generation of solvent gases and different contractions in the material. The shrinkage of the piece during drying occurs due to the removal of the liquid between the particles of the shaped material. This should be as uniform as possible, since differences in the contraction generate stresses that produce cracks in the material. The control in the drying of the green piece is done in controlled atmosphere and temperature chambers. In the organic elimination stage, the organic compounds used during the processing are progressively eliminated from the outside to the inside, so that the greater the thickness of the piece obtained, the more difficult it is to eliminate it. The processes of elimination of organic additives are: evaporation, chemical extraction with solvents, catalytic extraction, or thermal decomposition.

Finally the sintering process takes place, in which the compacted material, in green, is subjected to a sufficiently high temperature so that the phenomena of mass transport to the interstices between particles take place and compact densification occurs. As the sintering progresses, the porous phase becomes discontinuous and is eliminated, leaving only the remaining pores occluded in the matrix of ceramic material and that cannot be eliminated. The initial density of the piece in green is usually 40–60% of the theoretical density, while the final density is 85–100%. Sintering can take place through three mechanisms: sintering in solid state, sintering in the presence of a liquid phase, and reactive sintering.

On the other hand, due to the fact that ceramic materials do not corrode, their useful life periods are often much greater than those of other materials. However, the same physical properties that make ceramics so durable also make them extremely difficult to recycle, with the exception of glass. Although recycling is difficult, there are ways to "revalue" them, that is, new uses can be given to certain waste. Thus, ceramic materials (bricks), concrete and bituminous pavements, can be used for the manufacture of secondary products, such as the production of aggregates, which in turn can be used to manufacture concrete or serve as direct bases for road works.

In addition, the ceramic or cement industry has manufacturing processes that make the recovery of waste especially feasible [7–9], either taking advantage of the calorific power coming from the combustion or incorporating the waste into the internal structure of the materials, forming part of its own matrix and becoming an inert element.

Therefore, it can be concluded that ceramic materials have a great versatility of applications, because they have very characteristic properties that cannot be obtained from any other material.

Author details

Dolores Eliche-Quesada[1*], Luis Pérez-Villarejo[1] and Pedro José Sánchez-Soto[2]

1 Department of Chemical, Environmental and Materials Engineering, Universidad de Jaén, Jaén, Spain

2 Materials Science Institute of Sevilla (ICMS), Joint Center Spanish National Research Council (CSIC)-University of Sevilla, Sevilla, Spain

*Address all correspondence to: deliche@ujaen.es

References

[1] Hare. Ceramics. Properties and applications. In: Martyn Grayson, editor. Encyclopedia of Glass, Ceramics and Cement. John Willey and Sons; 1984

[2] Freiman W. Introduction to ceramic and glasses. In: Engineered Materials Handbook, Ceramic and Glasses. Vol. 4. ASM International; 1991

[3] Cranmer C. Overview of technical, engineering and advanced ceramics. In: Engineered Materials Handbook, Ceramic and Glasses. Vol. 4. ASM International; 1991

[4] Borrel-Tomás MA, Salvador-Moya MD. Advanced ceramic materials: Properties and applications. Universidad Politécnica de Valencia; 2018

[5] Barry Carter C, Grant Norton M. Ceramic Materials, Science and Engineering. New York, NY: Springer; 2013

[6] Rahaman MN. Ceramic Processing. Taylor & Francis Group; 2007

[7] Pérez JA, Terradas R, Manent MR, Sejas M, Martinez S. Inertization of industrial wastes in ceramic materials. Indian Ceramics. 1996;**16**:7

[8] Pereira DA, Aguilar JB, Castro FP, Almeida MF, Labrincha JA. Mechanical behaviour of Portland cement mortars with incorporation of Al-rich salt slags. Cement and Concrete Research. 2000; **30**:1131

[9] Andreola F, Barbieri L, Lancellotti I, Leonelli C, Manfredini T. Recycling of industrial wastes in ceramic manufacturing: State of art and glass case studies. Ceramics International. 2016;**42**:13333-13338

Chapter 2

The Use of Ceramic Waste Powder (CWP) in Making Eco-Friendly Concretes

Amr S. El-Dieb, Mahmoud R. Taha and Samir I. Abu-Eishah

Abstract

The global production of ceramic waste powder (CWP), which is produced during the final polishing process of ceramic tiles, exceeds 22 billion tons. The disposal of CWP in landfills will cause significant environmental problems (i.e., soil, air, and groundwater pollution). CWP is characterized by its chemical composition that is mainly composed of silica (SiO_2) and alumina (Al_2O_3). Both minerals represent more than 80% of the CWP composition. CWP has potentials to be used as an ingredient to partially or entirely replacing Portland cement to make eco-friendly concretes. This chapter summarizes the effect of using CWP in making eco-friendly concretes, with a particular focus on using CWP as a partial cement replacement in conventional-vibrated concrete (CVC) and self-compacting concrete (SCC), and the production of zero-cement alkali-activated concrete (AAC).

Keywords: ceramic waste powder, cement replacement, eco-friendly concrete, durability, microstructure

1. Introduction

In a rapidly growing world population and toward meeting consumers' needs, solid waste landfills will continue receiving huge volumes of waste. Therefore, waste management is becoming increasingly mandatory for the promotion of environmental sustainability. Numerous regulations have been imposed worldwide by governments and environmental organizations in order to reduce the negative environmental impact resulting from large numbers of solid waste landfills. The transformation of a large amount of solid waste into an alternative resource will preserve the reducing nonrenewable resources of materials; maintain the required energy and also will help solve environmental and exhausted landfill problems. Until today, researchers are investigating new solid waste materials and the potentials of recycling either in other industries or new products.

Being the world's most consumed human-made material, concrete attracted considerable interest as a possible way to recycle solid waste products especially those that can replace cement which is a significant contributor to global greenhouse gas emissions. An equal amount of CO_2 is generated for the production of Portland cement [1]. The cement industry produces around 5–8% of the annual global greenhouse gas emissions released into the atmosphere [2]. Several by-products such as fly ash, slag, and silica fume are effectively being used in the daily production of concrete as partial cement replacement (i.e., supplementary cementitious materials (SCM)) to reduce CO_2 emission [3, 4].

Global production of ceramic tiles is more than 12 Billion m^2 [5]. The manufacture of ceramic tiles generates ceramic waste powder (CWP) during the final polishing process at a rate of 19 kg/m^2 [6]. Therefore, the global generation of CWP exceeds 22 Billion tons. The CWP represents a significant challenge to get rid of concerning its environmental impact. It can cause, soil, water, and air pollution. On the other hand, it could represent an excellent opportunity to be used as an alternative concrete ingredient if it could be utilized in making concrete.

The effect of using ceramic wastes (i.e., roof tiles, blocks, bricks, electrical insulators, etc.) as aggregates or SCM in conventional-vibrated concrete (CVC) and mortar was reported in several studies. It is noted that limited studies were conducted on using CWP as a cement replacement in self-compacting concrete (SCC) and alkali-activated concrete (AAC) (i.e., geopolymer concrete). Some studies investigated the use of ceramic waste as coarse aggregates in CVC and mortar [7–16]. It was concluded that ceramic waste could be used as partial replacement of natural coarse aggregate. The ceramic waste aggregate should be pre-saturated by water to offset its high absorption. The compressive strength decreased if the ceramic waste replaced natural coarse aggregate beyond 25% by weight. The use of ceramic waste as fine aggregate in CVC and mortar was assessed by various researchers [16–22]. It was noted that using a high content of ceramic waste as fine aggregate had a negative impact on the workability of the fresh concrete, and workability admixtures were needed to avoid any adverse effect on concrete workability. It was concluded that the use of 50% by weight replacement of fine natural aggregate by ceramic waste could produce concrete without affecting the performance of hardened concrete.

The use of CWP as partial replacement of cement attracted the attention of several researchers [6, 23–35]. The main conclusion from the studies was that CWP showed slow pozzolanic activity which was evidenced at late ages. The early compressive strength was reduced by the inclusion of CWP. The development of compressive strength needed time. On the other hand, durability was improved by the incorporation of CWP in the mixtures. It was noticed that the investigations on using CWP as partial replacement of cement did not address the fresh concrete properties as affected by the inclusion of CWP as well as the microstructure characteristics. Also, no guidelines were provided for using CWP to partially replace cement. The CWP replacement level will depend on personal knowledge and experience. Furthermore, the replacement of cement by large quantities of CWP needs further evaluation.

The use of CWP in self-compacting concrete (SCC) mixtures received limited attention. In 2017, Subaşi et al. [36] investigated the use of CWP as a partial cement replacement in SCC mixtures. It was concluded that CWP could replace 15% by weight of the cement without adversely affecting the properties of the produced SCC. In 2018, Jerônimo et al. [37] replaced cement by ground clay brick waste (GCBW) in SCC mixtures. It was concluded that 20–30% by weight of the cement could be replaced by GCBW, and the compressive strength improved at 90 days of age. It was observed that the detailed evaluation of the SCC fresh properties as affected by the inclusion of CWP was not addressed. Also, the effect of using high-volume CWP in SCC still needs further assessment.

Concerning using CWP in alkali-activated concrete (AAC) (i.e., geopolymer concrete), it was noted that very limited investigations were conducted [38–40]. The main conclusion that CWP could be used in making AAC but needs detailed investigation and assessment.

An in-depth investigation to study the utilization of CWP in the production of different types of concrete is needed. This chapter summarizes the findings of collective studies conducted by the authors investigating the use of CWP in making eco-friendly concrete [41–45], with a particular focus on using CWP as a partial cement replacement in CVC and SCC, and the production of AAC. This will establish better understanding on how to incorporate an existing solid waste as a new construction ingredient in making echo-friendly concretes in order to optimize solid waste management, and help protect the environment by reducing the use of cement and efficiently getting rid of a solid waste material.

2. Characteristics of CWP

The produced ceramic waste material was a wet material due to the use of water during the polishing process. The average moisture content was 36% by mass. The average specific surface area (SSA) measured by air-permeability (i.e., Blain air permeability test apparatus) was 555 m^2/kg. More than 50% by volume of the CWP particles had a size ranging between 5 and 10 μm. **Figure 1** shows the particles' size distribution of the CWP.

The CWP consisted of irregular and angular particles which are similar to cement particles in shape as shown in the scanning electron microscope (SEM) image in **Figure 2**. **Figure 3** shows the energy dispersive spectroscopy (EDS) of the main oxides of the CWP. The EDS analysis indicated that CWP is mainly composed of SiO_2 and Al_2O_3.

Table 1 gives the chemical analysis of the CWP as determined by X-ray fluorescence (XRF). CWP is mainly composed of silica (SiO_2) and alumina (Al_2O_3). Both oxides are around 85% of the total material mass. Other compounds (i.e., CaO, MgO, and SO_3) exist in small quantities. The mass fractions of (SiO_2 + Al_2O_3 + Fe_2O_3) satisfies the requirement of the ASTM C618 [46] for natural pozzolana (i.e., >70%). Also, the SO_3 and the loss on ignition (L.O.I.) conformed to the ASTM C618 requirements.

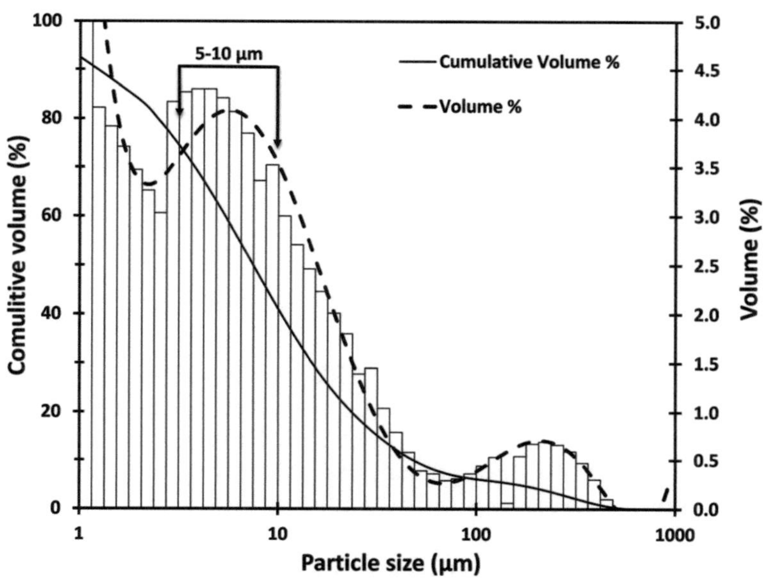

Figure 1.
Particle size distribution of CWP [43]. Reproduced with permission from the publisher.

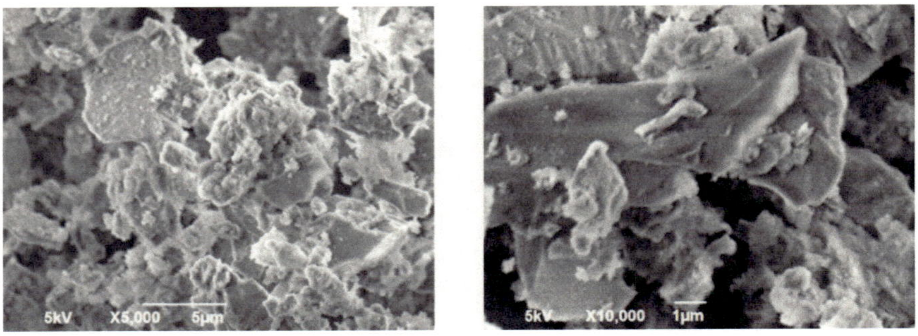

Figure 2.
SEM images of CWP.

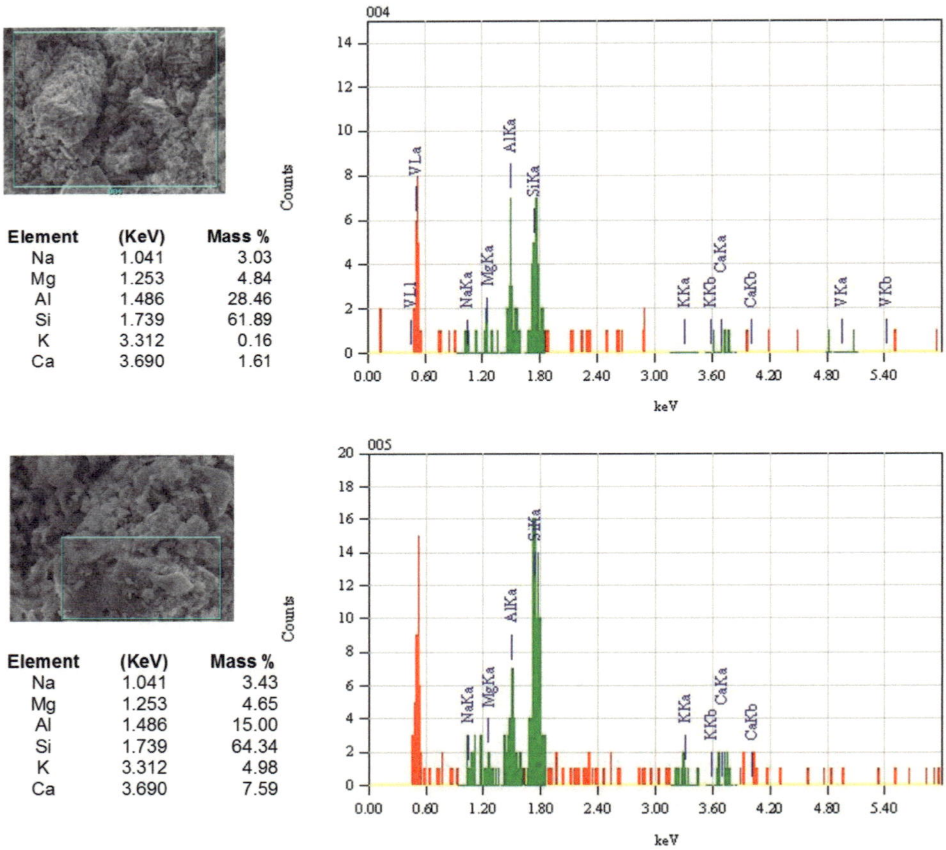

Figure 3.
EDS analysis of CWP [43]. Reproduced with permission from the publisher.

Figure 4 displays the X-ray diffraction (XRD) analysis of the CWP. The XRD indicates that the main peaks were noticed between 2-theta values of 20 and 30° which indicates the presence of (SiO_2). The observed hump between 20 and 30° indicates the occurrence of an amorphous phase. Moreover, the unleveled graph trend between the 2-theta values 0 and 40° indicates the existence of an amorphous phase in the CWP sample.

Characterizing industrial waste materials and their potentials is one of the challenging issues in the field of cement and concrete. The compressive strength was given prominence as an initial means for evaluating the pozzolanic activity. The compressive strength development of cement mortar including CWP is assessed according to ASTM C311 [47] to measure the strength activity index (SAI).

Four mortar mixtures are prepared in which cement is partially replaced by CWP. The replacement levels are 10, 20, 30 and 40% by weight. Strength activity index (SAI) is calculated as the strength percentage as compared to the control mortar mixture. **Table 2** gives the 28 days compressive strength, standard deviation SAI. Results showed that all CWP specimens satisfied the ASTM C618 requirement of SAI (i.e., >75%). In an investigation by Steiner et al. [25], a similar trend in the activity index for mortar mixtures with ceramic tiles polishing residues was reported. The SAI decreased after the inclusion of 40% CWP by cement mass; this could be attributed to the dilution effect. Also, it might be due to the high silica available in the mixture as a result of the high CWP. This large quantity could not find sufficient calcium hydroxide (CH) in order to react with. Therefore, most of the silica components were left without getting involved in the chemical reaction [48]. Also, Frattini test [49] is performed to identify the pozzolanic activity of CWP following BS EN 196-5:2011 [50]. Test samples with 0, 20 and 40% CWP as cement replacement by weight are tested. The Frattini test showed that concrete with 20

CaO	SiO_2	Al_2O_3	MgO	Fe_2O_3	SO_3	L.O.I.
1.70(0.69)	68.60(0.97)	17.00(0.57)	2.50(0.90)	0.80(0.04)	0.12(0.16)	1.78
Note: Values in parentheses are the standard deviation.						

Table 1.
Chemical composition of CWP using XRF (modified from [43]).

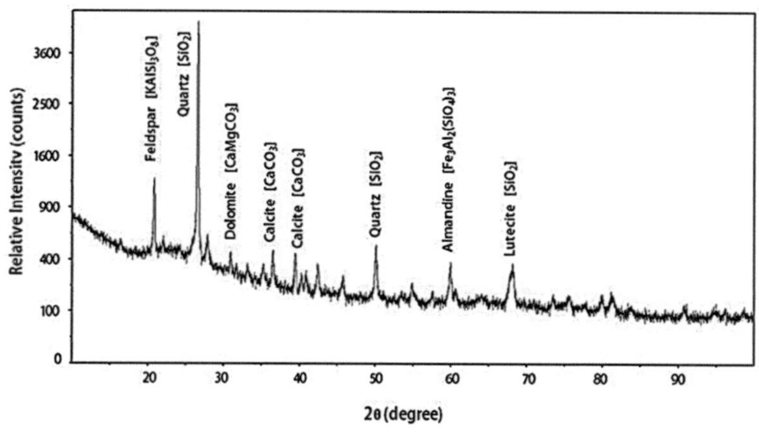

Figure 4.
XRD pattern of CWP [43]. Reproduced with permission from the publisher.

	CWP replacement level (mass %)			
	10%	20%	30%	40%
Average 28 days strength (MPa)	39.9	46.0	48.8	37.5
Standard deviation (MPa)	4.0	3.0	4.4	1.2
Strength activity index (SAI) in (%)	91.0	105.0	110.5	85.5
Reproduced with permission from the publisher.				

Table 2.
Strength activity index (SAI) results for CWP [43].

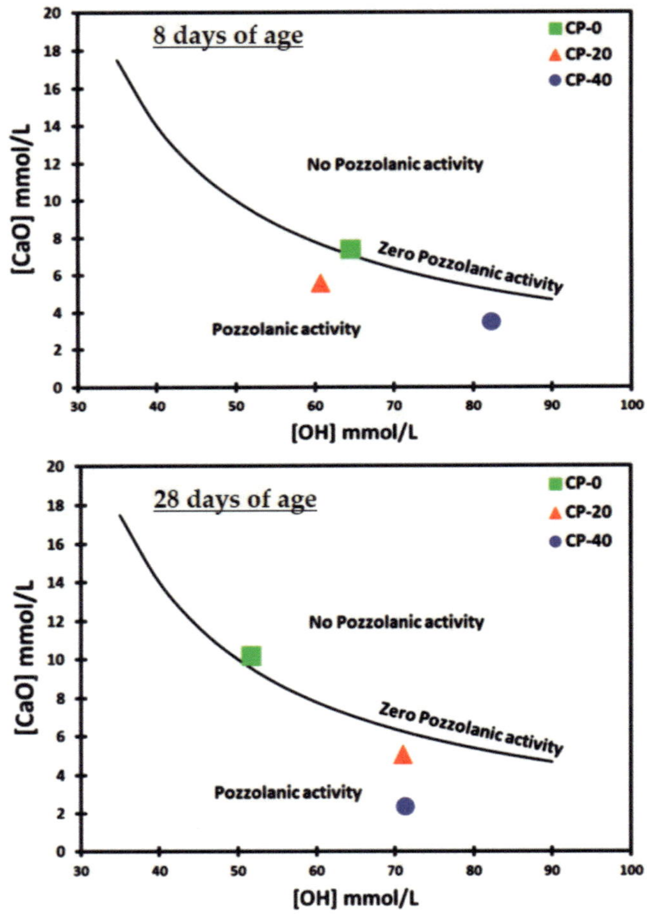

Figure 5.
Frattini test at 8 and 28 days of CP with CWP replacement [45]. Reproduced with permission from the publisher.

and 40% CWP replacement of Portland cement exhibited pozzolanic activity at 8 and 28 days age of concrete as shown in **Figure 5**.

In conclusion, CWP is silica and alumina rich material with some amorphous phases. The CWP has some pozzolanic activity, especially at a late age, as confirmed by strength activity index and Frattini tests. Therefore, CWP possesses the potentials to be used as a partial cement replacement in CVC and SCC mixtures, and as a main binder source to make AAC mixtures.

3. Conventional-vibrated concrete (CVC)

CWP is used to partially replace cement (0, 10, 20, 30 and 40% by weight) in different CVC mixtures. Two concrete grades with different cement contents are studied (25 and 50 MPa). The mixtures are chosen to cover several applications and different cement contents. All mixtures are designed to have a slump value from 60 to 100 mm. **Table 3** gives the mixtures' proportions of the mixtures. Initial slump values (i.e., ASTM C 143 [51]) is used to judge the mixtures' workability. The time to reach zero slump is used to assess the workability retention of the concrete mixtures. The development of compressive strength with age (i.e., 7, 28 and 90 days) and drying shrinkage (i.e., 120 days) are measured. Rapid chloride ion penetration

Mixture I.D.	Cement	CWP	Fine aggregate	Coarse aggregate	Water content	Initial slump (mm)
M25-0	310	0	749	1102	190	110
M25-10	279	31	737	1105	190	130
M25-20	248	62	734	1101	190	103
M25-30	217	93	731	1097	190	95
M25-40	186	124	629	1093	190	55
M50-0	485	0	662	993	208	55
M50-10	437	48	658	988	208	65
M50-20	388	97	654	981	208	60
M50-30	340	145	650	975	208	42
M50-40	291	194	673	968	208	10

Table 3.
Mixtures' proportions (kg/m^3) and initial slump values (mm) (modified from [43]).

test (RCPT) (i.e., ASTM C 1202 [52]) and bulk electrical resistivity test (i.e., ASTM C 1760 [53]) are conducted at 28 and 90 days of age to evaluate the durability of the concrete mixtures. Triplicate samples are used for the compressive strength, drying shrinkage, RCPT, bulk electrical resistivity and permeable pores tests and the average results are used. The development of the microstructure is assessed by measuring permeable pores (i.e., ASTM C642 [54]) and the pore system (i.e., total porosity and median pore diameter) is measured by mercury intrusion porosimetry (MIP). Both are measured at 90 days of age. Main microstructure characteristics are identified using scanning electron microscopy (SEM).

Concrete mixtures are prepared using ordinary Portland cement (OPC) as the primary binder. The specific surface area of cement is 380 m^2/kg. Natural crushed stone of maximum size 19.0 mm is used as coarse aggregate. The specific gravity is 2.65 while the absorption was 1%. Natural sand with fineness modulus between 2.5 and 2.7 is used as fine aggregate. The specific gravity is 2.63.

3.1 Workability and workability retention of fresh concrete

Initial slump values are given in **Table 3**. As CWP inclusion level increases, the initial slump value decreases as a result of its high specific surface area (SSA) compared to that of the cement (i.e., the SSA of CWP is 1.5 times that of the cement). Workability retention defines the time available for easy handling the mixture. **Figure 6** shows the time to zero slump of the concrete mixtures including CWP. It is noted that the workability retention time increases due to the inclusion of CWP. This could a result of CWP has no hydraulic reaction, and its pozzolanic reaction is slow. The use of 10% CWP in the 25 MPa mixtures has the highest workability retention. While for the 50 MPa mixtures, the use of 20% CWP shows the best retention time.

3.2 Compressive strength

The compressive strength development at different ages is shown in **Figure 7**. The coefficient of variation (COV) ranged from 0.4 to 4.8%. The compressive strength values at 7 and 28 days of age are lower than the target strength for both mixtures (i.e., 25 and 50 MPa). The reduction in strength is proportional to the CWP content.

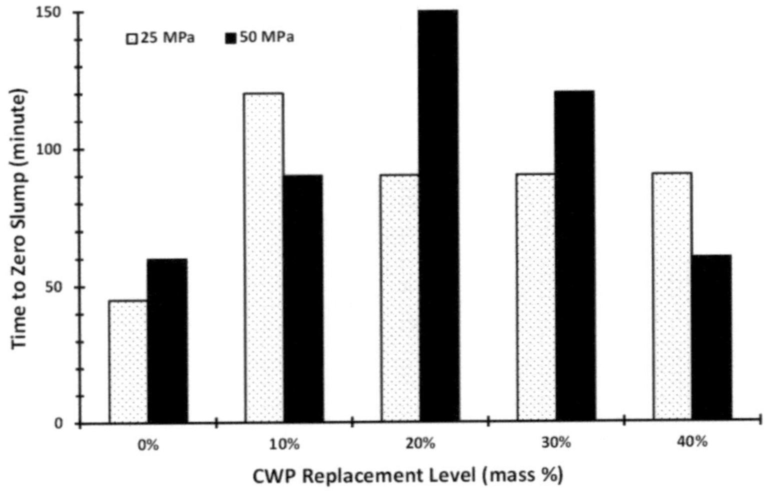

Figure 6.
Time to zero slump.

This could be attributed to the fact that CWP has no hydraulic reaction. Also, its contribution to early strength depended mainly on its microfilling ability (i.e., CWP particles' size ranged from 5 to 10 μm). This behavior agrees with that of most pozzolanic materials with slow strength development at early ages [55]. Also, slowed strength development at early ages is reported for CWP [28–30, 32].

At a late age (i.e., 90 days) all the 25 MPa mixtures including CWP achieve compressive strength values higher than the target strength. The mixture with 10% CWP shows the highest compressive strength. The strength gain at 90 days of age might be due to the pozzolanic characteristics of the CWP material. For the 50 MPa mixtures, all CWP mixtures the target strength is achieved. The increase in strength values could be justified by the delayed pozzolanic reaction of the CWP. The CWP particles could have worked as nucleation sites for cement grains and hydration products which led to a denser microstructure.

3.3 Drying shrinkage

Table 4 shows the 120 days drying shrinkage strain values. The COV ranged from 20 to 26%. It is observed that the drying shrinkage strain decreases with increasing the CWP replacement level. The pores' structure and connectivity of pores are changed due to the fine CWP particles and its pozzolanic action. This change results in restricting water movement through the concrete. The drying shrinkage values for mixtures including 10 and 20% CWP do not differ significantly from that of the control mixtures. For the 25 MPa mixtures, CWP with replacement levels of more than 20% reduces the drying shrinkage strain between 29 and 60% compared to the control mixture. While for the 50 MPa mixtures a decrease in the drying shrinkage strain values between 28 and 53% for CWP replacement levels above 20% are observed.

3.4 Chloride ion penetration test (RCPT)

The concrete durability concerning its resistance to chloride ion penetration and chloride induced corrosion can be judged by the RCPT. The inclusion of CWP as partial cement replacement has a significant effect on the chloride ion penetration

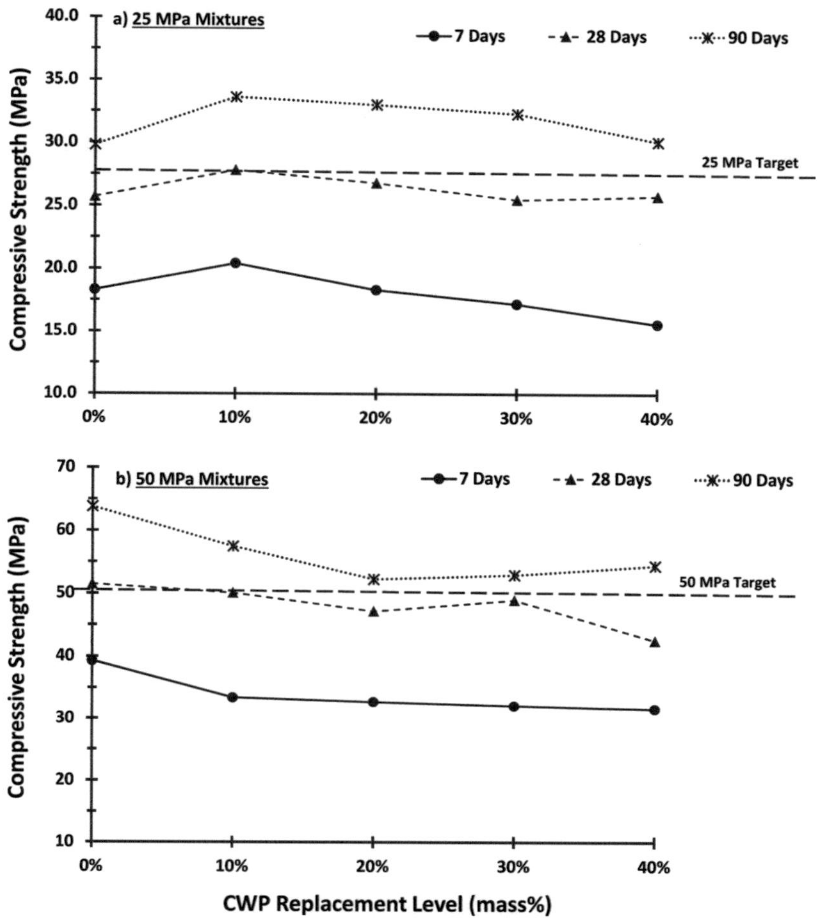

Figure 7.
Compressive strength development with age.

Mixture	Shrinkage strain (microstrain)	Mixture	Shrinkage strain (microstrain)
M25-0	2608	M50-0	2569
M25-10	2488	M50-10	2222
M25-20	2817	M50-20	2413
M25-30	1033	M50-30	1199
M25-40	1859	M50-40	1848

Table 4.
Drying shrinkage strain values at 120 days (microstrain) (modified from [43]).

of the 25 and 50 MPa concrete mixtures. **Figure 8** demonstrates a significant reduction in the 28 and 90 days' test results of all CWP concrete mixtures. The COV ranged from 3 to 15%.

At 28 days of age, the use of 20, 30 and 40% CWP reduces the total passed charge by 2–8 times lower than that of the control mixture. Mixtures with 30 and 40% are rated as "Very Low" for chloride ion penetration as per the classification of the ASTM C1202 [52]. At 90 days of age, the chloride ion penetration classification of all the 25 MPa mixtures including CWP is "Very low." The reduction in the total

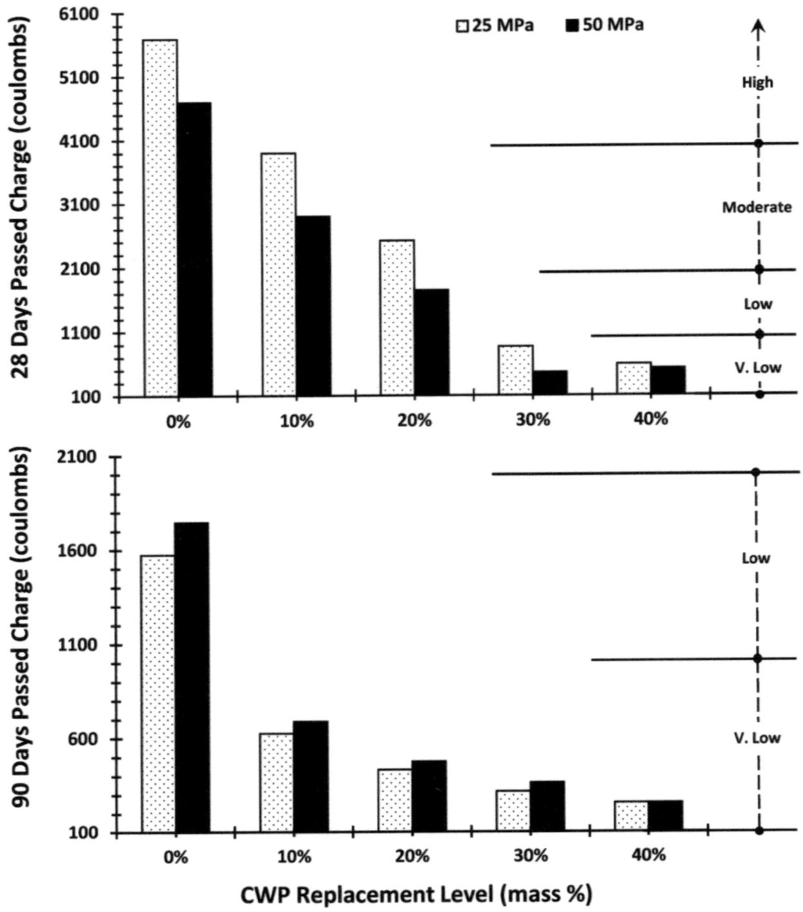

Figure 8.
Chloride ion penetration.

passed charge for the mixtures incorporating CWP compared to its corresponding 28 days values ranged from 56 to 84%.

While for the 50 MPa mixtures, the 28 days chloride ion penetration decreases with the inclusion of CWP. The reduction is proportional to the CWP content. The reduction with respect to the control mixture is 38% for the use of 10% CWP and 90% for the use of 40% CWP. The ASTM classification of mixtures including high levels of CWP (i.e., ≥20) is shifted from "High" to "Low" and even "Very Low." At the 90 days of age, chloride ion penetration for all 50 MPa CWP mixtures is classified as "Very Low." This significant reduction could be due to the microstructure densification and refinement of the pore structure provided by the fine particles of CWP in addition to its pozzolanic effect. Also, the reduction with age indicates the development of a dense microstructure, especially with discontinuous pore system. Similar findings were reported in other studies [6, 30, 34, 56].

3.5 Bulk electrical resistivity test

The corrosion protection of the concrete to the embedded reinforcement can be assessed by its electrical resistivity [57]. **Figure 9** displays the bulk electrical resistivity at 28 and 90 days of age. The COV ranged from 4 to 10%. It should be noted that electrical resistivity is mainly affected by the porosity and the pore size distribution [58]. Therefore, the development of the microstructure could be judged by measuring the electrical resistivity. Ionic mobility is reduced by the discontinuity of pores, and hence concrete resistivity and corrosion protection will increase. The resistivity results of all concrete mixtures including CWP are higher than those of the control mixtures. Microfilling effect and pozzolanic activity of the CWP which could lead to a denser microstructure could be the main reasons for the increase in the resistivity of the mixtures including CWP. It was reported that the use of ceramic polishing residues was reported to reduce water permeability of cement mortar samples [6, 34].

At 28 days of age, 25 MPa mixtures including 20, 30 and 40% CWP have a resistivity higher than 10 kΩ.cm. This is classified as "High" to "Very High" corrosion protection levels according to ACI 222R-01 [57]. The increase in resistivity is proportional to the CWP replacement level. At 90 days of age, using CWP demonstrates a significant increase in the electrical resistivity values with respect to the

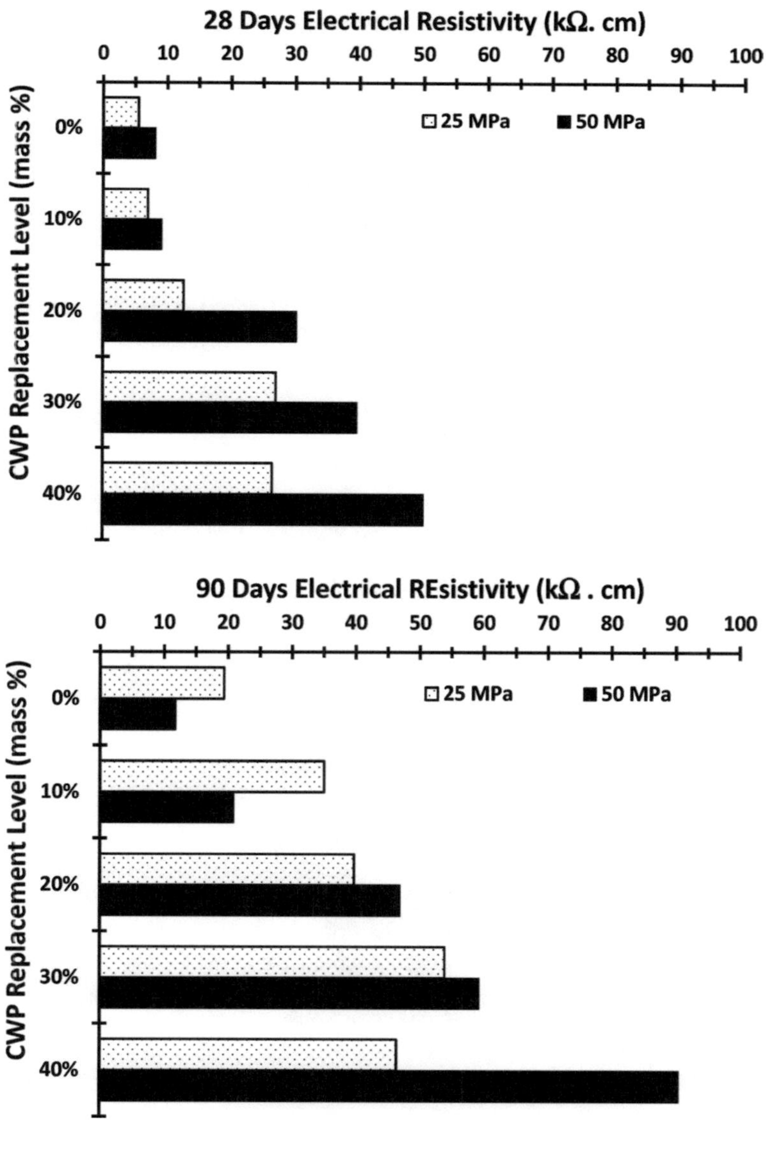

Figure 9.
Bulk electrical resistivity.

control mixture. The 50 MPa concrete mixtures with CWP had similar performance to the 25 MPa mixtures at both ages. Including 10% CWP results in a "High" corrosion protection level. When CWP is included with 20% or more the corrosion protection level is "Very High" at both ages.

Both RCPT and resistivity results confirm the performance of the concrete mixtures including CWP with regards to chloride ion attack, chloride-induced corrosion, and corrosion protection.

3.6 Permeable pores

The permeable pores of the concrete mixtures can assess the development of the pore system and judge the microstructure development. **Figure 10** shows the permeable pores measured at 90 days of age. The COV ranged from 2 to 8%. In general, the permeable pores are decreased by the inclusion of CWP compared to the control mixture.

In the case of the 25 MPa mixtures, the permeable pores are reduced by 17% up to 36% due to the inclusion of CWP as a partial cement replacement. Similar performance is observed for the 50 MPa mixtures. The reduction in pores volume ranged from 2 to 24% compared to the control mixture. The inclusion of the fine CWP particles with high SSA could physically have a microfilling effect and improves the particles' packing in the mixtures. Also, to the CWP pozzolanic activity, the mixtures microstructure is densified. Therefore, the pore structure is refined resulting in lower pore volume. The reduction in permeable pores reduces the mobility of water from inside the concrete which is reflected in reducing the reduction in the drying shrinkage strain. Also, reduction in chloride ion penetration and immobility of ions are direct effects of the pores' size refinement. This is reflected in the reduction of the chloride ions penetration and the improvement of the electrical resistivity with age.

3.7 Mercury intrusion porosimetry (MIP)

MIP is a widely used test to characterize the pore structure of cement-based materials. The test is capable of providing information about the total porosity, and the median pore diameter based on intruded volume. The concrete pore system indicates its microstructural development that can be related to its performance.

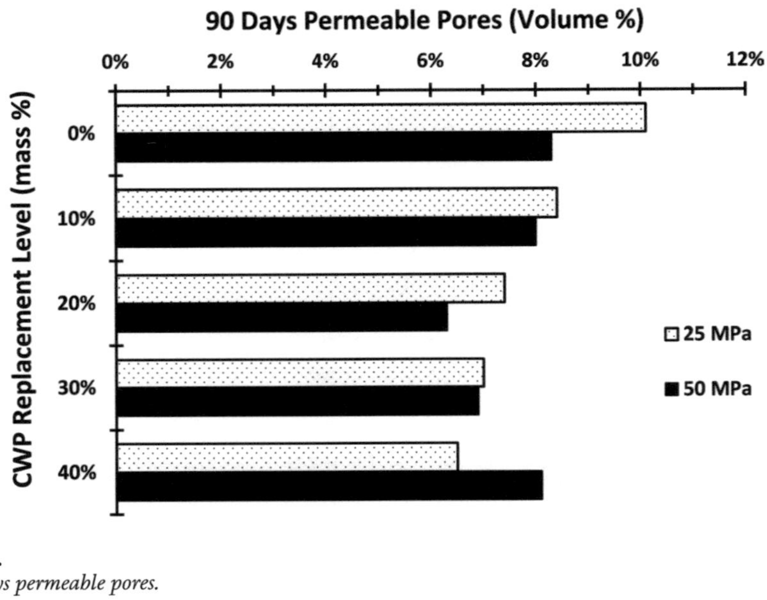

Figure 10.
Ninety days permeable pores.

Table 5 gives the results of the MIP test regarding total porosity percentage and the median pore diameter based on intruded volume at 90 days of age. The inclusion of CWP reduces the total porosity at 90 days of age. The use of 40% CWP as partial replacement of the cement reduces the porosity by 9 and 19% for the 25 and 50 MPa mixtures respectively compared to the same mixtures without CWP. The median pore diameter is reduced due to the inclusion of CWP. It is noted that the reduction was proportional to the CWP content. The reduction in the total porosity and the median pore diameter confirms the densification of the microstructure due to the inclusion of CWP as a partial cement replacement.

The reduction in the total porosity and especially the reduction in the pore size confirm the superior durability performance of the mixture observed at the late age. The microstructure development could be related to the durability performance. The median pore diameter was correlated to the 90 days RCPT and electrical resistivity values as shown in **Figure 11**. The median pore diameter correlates well with the durability test results. The correlation coefficient (R^2) is 0.9517 and 0.7977 for the median pore diameter relationship with the RCPT and the electrical resistivity respectively.

3.8 Microstructure characteristics

To better understand the performance of CVC mixtures including CWP, the main microstructural characteristics are inspected by scanning electron microscope (SEM). Microstructure examination is conducted at 90 days of age. The examination is conducted on the control mixture for both concrete grades (i.e., M25-0 and M50-0), and the mixtures including the highest CWP content (i.e., M25-40 and M50-40).

Figure 12 shows the SEM images of the general characteristics for M25-0 and M25-40. For the M25-0 mixture, crystalline hydration products are observed in addition to several pores. For M25-40, fewer pores with smaller size are noticed which indicates the densification of the microstructure that confirms the superior durability performance. Few crystalline hydration products are observed. **Figure 13** displays the aggregate matrix interfacial transition zone (ITZ) for M25-0 and M25-40 mixtures. Crystalline hydration products are noticed in both mixtures in the ITZ region with smaller crystal size in M25-40 mixture. The matrix around the aggregate in the M25-40 mixture includes lesser pores compared to M25-0, this is similar to the observations of the general matrix microstructure.

Mixture	Porosity (%)	Median pore diameter* (μm)
M25-0	21.297	4.2586
M25-10	20.015	4.0115
M25-20	19.754	3.7404
M25-30	19.135	3.6184
M25-40	19.437	3.4737
M50-0	22.426	4.0380
M50-10	21.131	3.8382
M50-20	19.415	3.5876
M50-30	18.944	3.5747
M50-40	18.126	3.4000

**Based on the intruded volume.*

Table 5.
MIP results at 90 days of age.

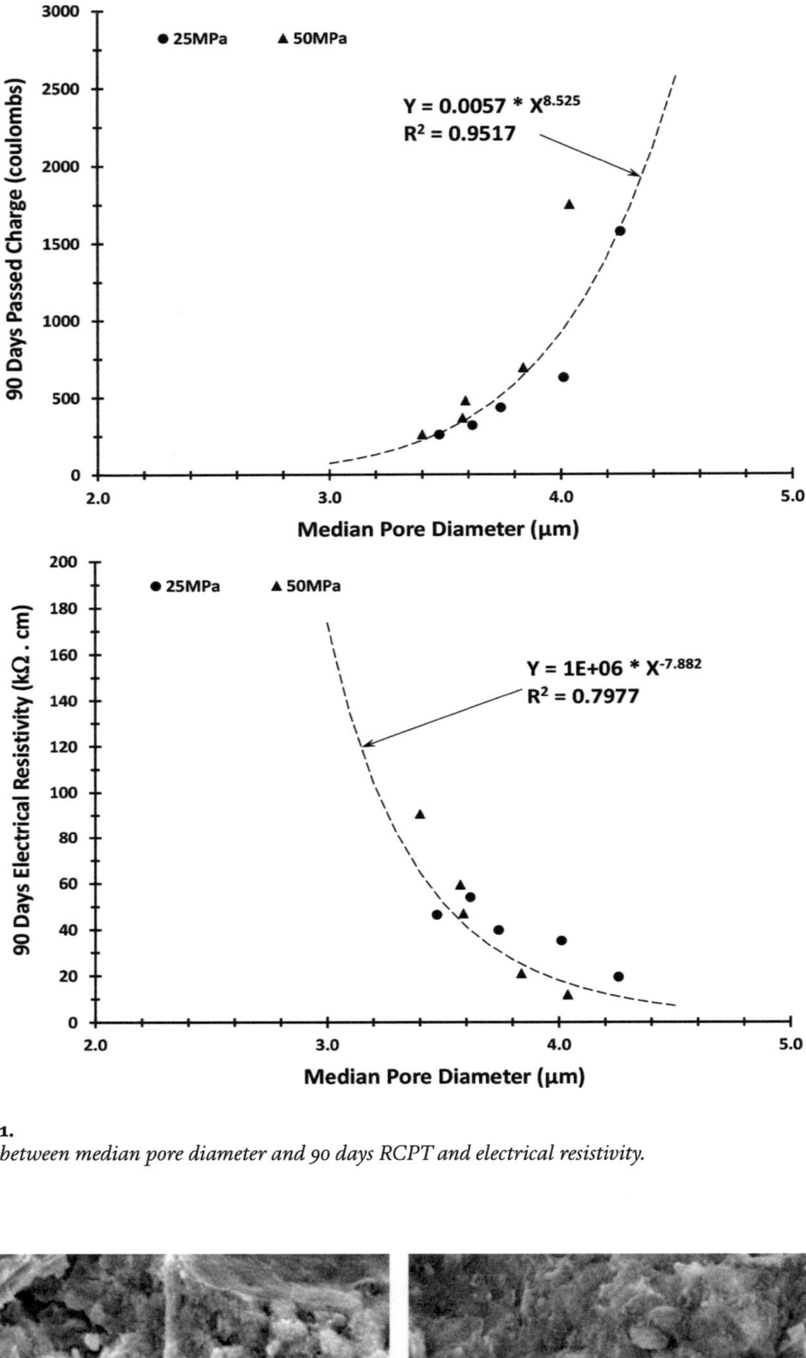

Figure 11.
Relation between median pore diameter and 90 days RCPT and electrical resistivity.

Figure 12.
SEM image of general microstructure for M25-0 and M25-40 mixtures.

The general microstructure for M50-0 and M50-40 is shown in **Figure 14**. Generally, the 50 MPa mixtures have a denser microstructure compared to the 25 MPa mixtures. For the M50-0 mixture, few pores are noticed, and the crystalline hydration products are smaller in size. The inclusion of CWP densified the microstructure by refining the pore structure as depicted in the SEM image. The ITZ region microstructure is presented in **Figure 15**. The incorporation of CWP improves the densification of the ITZ region microstructure. The crystalline hydration products and pores' size are reduced due to the inclusion of CWP.

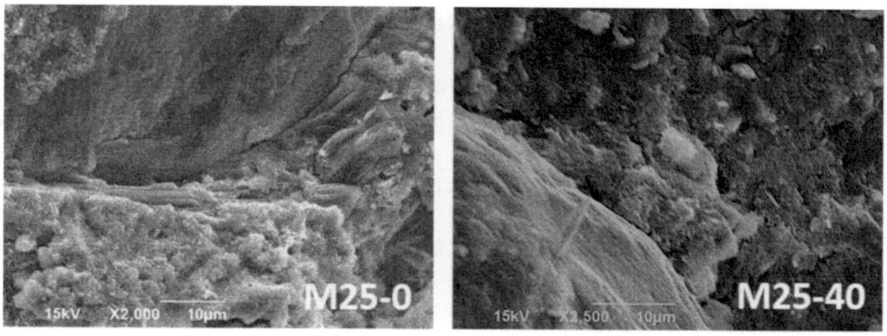

Figure 13.
SEM image of ITZ region for M25-0 and M25-40 mixtures.

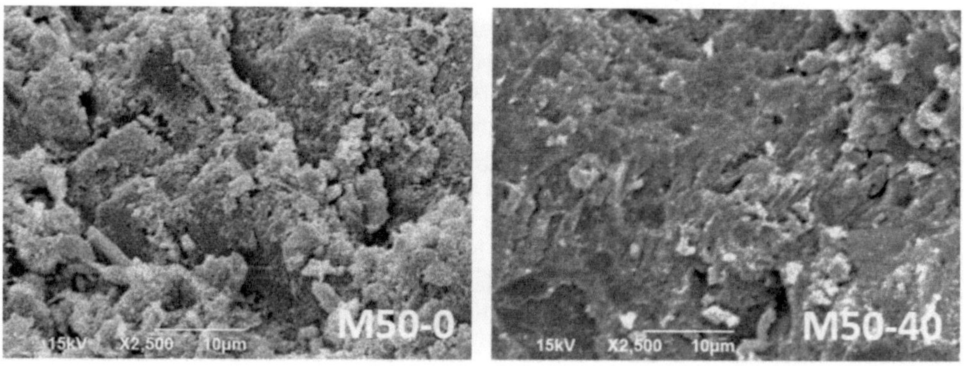

Figure 14.
SEM image of general microstructure for M50-0 and M50-40 mixtures.

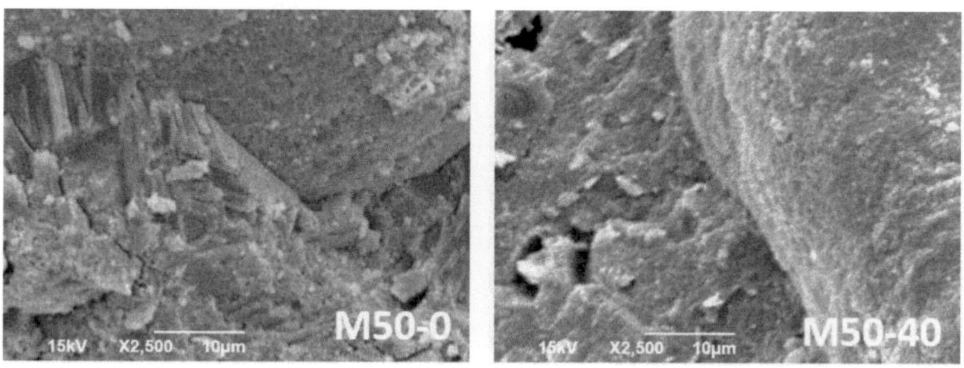

Figure 15.
SEM image of ITZ region for M50-0 and M50-40 mixtures.

4. Self-compacting concrete (SCC)

Self-compacting concrete (SCC) has received wide attention and used in the construction industry worldwide since its development [59]. SCC is featured with high fluidity, and at the same time, high resistance to segregation and is placed purely under its weight without the need for vibration [60–62]. SCC properties are the result of modifying the composition of CVC by incorporating high powder content that has been mainly cement. However, the use of high cement content is not desirable as it will increase the cost and has other negative environmental effects. Replacing cement in SCC mixtures with waste powder is a trend gaining a great deal of attention with the growing awareness toward environmental protection and sustainable construction [63–70]. CWP is used to partially replace cement to produce eco-friendly SCC. The cement content in the control mixture is 500 kg/m^3 based on the preliminary mix design. The powder content of the control mixture meets the recommended value by EFNARC specifications [71]. The cement is partially replaced by the CWP in 20, 40 and 60% by weight. The concrete mixture is expected to yield compressive strength in the range of 80 MPa. The details of the mixtures' proportions are given in **Table 6**.

Ordinary Portland cement (OPC) is used as the main binder. The specific surface area of cement is 380 m^2/kg. Natural crushed stone of maximum size 9.5 mm is used as coarse aggregate. The specific gravity is 2.65 while the absorption was 0.7%. Natural sand with fineness modulus between 2.5 and 2.7 is used as fine aggregate. The specific gravity is 2.63.

Several tests are conducted to investigate the effect of replacing cement with CWP on the fresh properties of the produced concrete. Unconfined flowability of the produced SCC mixture is assessed by the slump flow test in accordance to ASTM C1611 [72]. Passing ability is evaluated through two tests namely the J-ring (i.e., ASTM C1621 [73]), and L-box. The segregation resistance is measured through conducting the GTM segregation column test conforming to ASTM C1610 [74]. Finally, the viscosity is measured by following the V-funnel test procedure described in the EFNARC specification [71]. On the other hand, compressive strength is performed at two test ages (i.e., 7 and 28 days) in order to evaluate the strength development. The durability characteristic is evaluated by conducting the bulk electrical resistivity as per ASTM C1760 [53] at 28 and 90 days of age. Triplicate samples are used to

Mixture ingredients	Mixture designation			
	Control	R-20	R-40	R-60
Cement	500	400	300	200
CWP	0	100	200	300
Water	175	175	175	175
Fine aggregate	871	871	871	871
Coarse aggregate	871	871	871	871
Super plasticizer	8.33	8.72	8.33	8.80
VMA*	1.6	1.6	1.6	1.6
w/cm**	0.35	0.35	0.35	0.35

**VMA = viscosity-modifying admixture.*
***w/cm = water/(cement + slag or CWP).*

Table 6.
Mixtures' proportions for SCC (kg/m^3).

conduct the compressive strength and the bulk electrical resistivity tests and the average results are used. **Figure 16** shows the different tests conducted. The microstructure development is judged by measuring the permeable pore volume at 28 and 90 days of age. Also, the pore system (i.e., total porosity and median pore diameter) is assessed using mercury intrusion porosimetry (MIP). The MIP is conducted at 90 days of age.

4.1 Slump flow results

Slump flow test evaluated the unconfined flowability of the produced SCC mixtures. **Figure 17** displays the test results together with the EFNARC specifications [71].

It is noticed that the slump flow decreases as the amount of CWP in the mixture increases. Even with the reduction in the slump flow values, none of the CWP

Figure 16.
Different tests conducted on SCC.

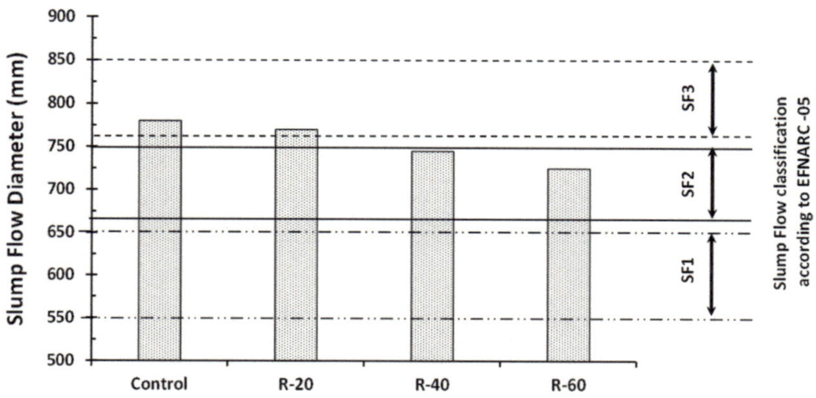

Figure 17.
Slump flow results.

mixtures dropped to the slump flow class one (SF1) which is critical in the presence of highly congested reinforced concrete structures.

Chopra and Siddique [48] reported a similar trend when using rice husk ash (RHA) as cement replacement. The relatively higher specific surface area (SSA) of the CWP compared with cement would increase the water demand and accordingly resulted in lower slump flow values. Similarly, Sfikas et al. [75] reported a reduction in the slump flow of SCC when they used metakaolin, which is characterized by a high SSA, to replace cement.

The time taken for concrete to reach the 500 mm diameter circle on the steel base plate of the slump flow test is measured (T50). The T50 value can judge the viscosity of the SCC mixtures. High T50 values indicate mixtures with higher viscosity. The T50 results are given in **Table 7**.

4.2 J-ring results

The passing ability of SCC is evaluated by the J-ring test. This test evaluates how the SCC mixtures can perform in the presence of reinforcing bars in form works. The difference between the unrestricted slump flow diameter and the J-ring flow diameter is shown in **Figure 18**. The inclusion of CWP improves the passing ability of the SCC mixtures. As the CWP content increases the mixtures' passing ability is improved and shows a great capacity for flowing through congested spaces. Therefore, mixtures containing high CWP perform better than the control mixture with regards to the passing ability.

4.3 L-box results

The passing ability of SCC through congested reinforcement can also be assessed by using the L-box test. The L-box results are given in **Table 7**. Comparable blocking ratios are observed for all tested mixtures. The variation is less than 1.5%. SCC mixtures including CWP mixtures show no signs of blocking. Generally, EFNARC [71] suggests blocking risk is likely if the blocking ratio is below 0.8. The viscosity of the mixtures is too high if the blocking ratio is less than 0.8. This can cause blocking around highly congested sections. Based on the results, all mixtures with CWP can be used in applications where flow through congested reinforcement is needed.

4.4 V-funnel results

In this test, the viscosity and filling ability of the fresh concrete is judged by the V-funnel test where the concrete is forced to flow through small cross sections and confined spaces. The flow rate (i.e., V-funnel time) of the SCC through the small cross-section is directly related to the mixture's viscosity.

The V-funnel test results are given in **Table 7**. The V-funnel results show an increasing trend, indicating a higher viscosity of the mixtures. All the measured

Property measured	Control	R-20	R-40	R-60
T50 (seconds)	2.68	2.47	3.24	4.04
V-Funnel (seconds)	10.4	10.01	11	12.82
L-box ratio (H2/H1)	0.963	0.966	0.977	0.967

Table 7.
Fresh test results (modified from [42]).

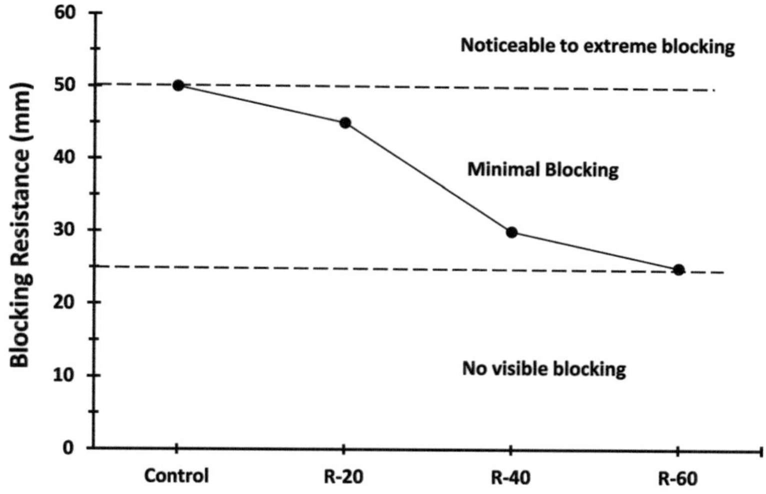

Figure 18.
J-ring results.

V-funnel time values correspond to the second viscosity class according to EFNARC specification [71]. The increase in the viscosity indicates an improvement in the segregation resistance. The viscosity-modifying admixture (VMA) is typically used to adjust mixtures' viscosity and enhance segregation resistance. Since the mixtures' viscosity values are significantly enhanced by the incorporation of CWP the VMA could be eliminated from the mixture or its dosage could be reduced. This would result in more economical and low-cost mixtures.

4.5 GTM segregation column results

The ability of concrete to remain homogeneous in the composition in its fresh state is defined as its segregation resistance. The GTM segregation column test is used to evaluate the mixtures' segregation resistance.

Segregation percentage is shown in **Figure 19**. The segregation percentage decreases as the CWP content increases in the mixtures. The CWP significantly improves the segregation resistance of the SCC mixtures. The incorporation of CWP in SCC enhances the cohesiveness characteristics of the mixtures. The segregation percentages are below 15%, which shows that the SCC mixtures were superior regarding segregation resistance. Segregation resistance is related to viscosity. The improvement in segregation resistance is confirmed by the V-funnel test results. As the amount of CWP increases in the mixtures from 0 to 60%, the segregation resistance is enhanced by 72.5%. The substantial enhancement in the segregation resistance can be explained by the fact that the water adsorption of the CWP particles may induce suction forces possibly leading to cluster formation. This will lead to an increase in the inter-particle bonds as in the clustering theory enhancing the segregation resistance similar to RHA mixtures studied by Le and Ludwig [76].

4.6 Compressive strength results

Strength is measured at different test ages (7, 28, and 90 days) to evaluate the strength development as affected by the inclusion of CWP as partial cement replacement. The strength development due to the inclusion of any cement replacing material is mainly affected by the cement hydration and pozzolanic reaction the used material, and the effect on the concrete microstructure especially the densification of the microstructure with a particular focus on the aggregate-paste interfacial zone [77].

Figure 20 shows the compressive strength development with age. The COV ranged from 0.4 to 3.0%. At the 7 days of age, it is noticed that the inclusion of CWP decreases the strength and the reduction is proportional to the CWP content. This could be a direct result of replacing cement by CWP which has no hydraulic reaction. At the 28 days of age, the mixture including 20% by weight CWP showed higher strength compared to the control mixture. Nevertheless, the mixture of 60% by weight CWP shows the least developed strength. Since CWP is characterized by the slow pozzolanic reaction, it is expected not to see much effect until late ages. At the 90 days of age, the improvement in strength is noticeable. At the 90 days of age, mixtures with 20 and 40% by weight CWP achieve the highest compressive strength compared to the control mixture. This implies that 20–40% by weight CWP is the optimum cement replacement to obtain high compressive strength.

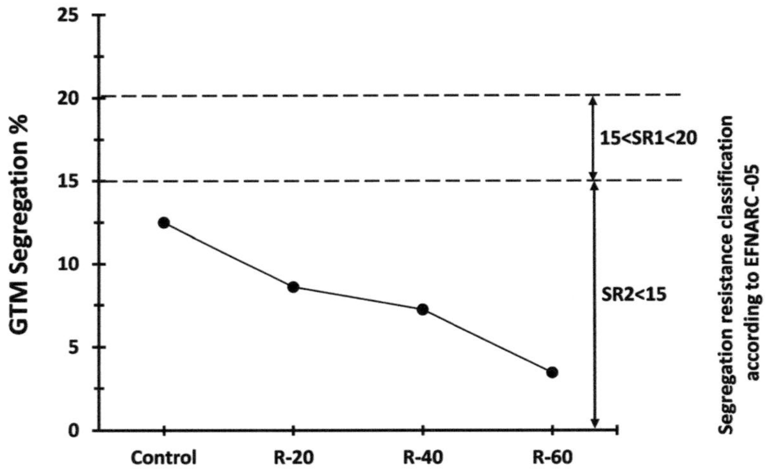

Figure 19.
Segregation resistance results.

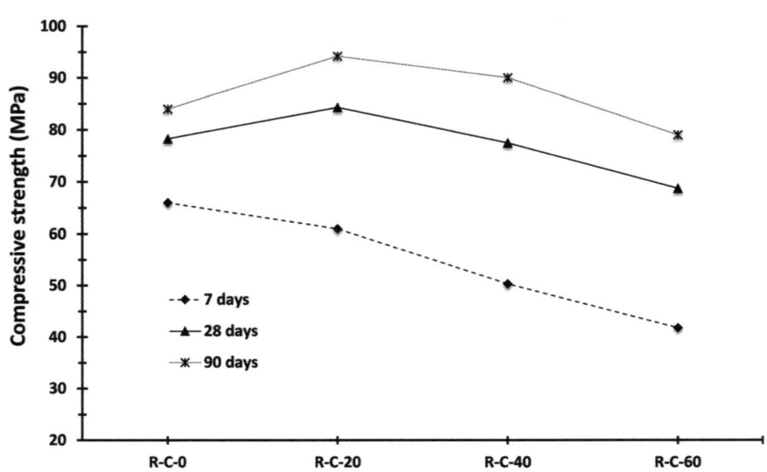

Figure 20.
Compressive strength development with age.

The increase in the strength can also be explained through the nucleation sites (i.e., nucleation of CH around the CWP particles). The CWP improves the packing of the concrete mixture due to its high SSA and its pozzolanic reaction, and the cement hydration acceleration similar to the effect of rice husk ash (RHA) observed in another investigation [76]. On the other hand, the use of 60% by weight CWP shows marginal improvement in strength; this can be due to the high amount of silica from the CWP, and the insufficient amount of calcium hydroxide (CH) from the cement hydration. Hence, some silica is left without chemical reaction. Similar behavior was observed by using RHA (i.e., characterized by high SSA and high silica content) as cement replacement [48].

4.7 Bulk electrical resistivity results

The electrical resistivity of concrete is affected by several factors such as porosity, pore size distribution, connectivity, concrete's moisture content, and ionic mobility in pore solution. Electrical resistivity assesses the concrete protection of reinforcing steel against corrosion. According to ACI 222R-01 [57], the corrosion protection level is improved as the resistivity value increases.

The resistivity values are presented in **Figure 21** at 28 and 90 days of age. The COV ranged from 6.4 to 13.2%. The resistivity increases with age. The inclusion of CWP significantly increases the mixtures' resistivity. The significant increase in the resistivity due to the inclusion of CWP suggests that CWP tended to reduce the interconnected pore network contributing to the reduction of the concrete's conductivity. With age, CWP pozzolanic activity contributes to the refinement of concrete pores and microstructure, thus further reduces the ionic mobility and hence the concrete's conductivity. The improved resistivity indicated that the durability of the CWP concrete mixtures to protect reinforcing steel against the corrosive environment is much better than that of the control mixture without CWP.

4.8 Mercury intrusion porosimetry (MIP)

The MIP test provides information about the pore system (i.e., pore volume and median pore diameter). The MIP results can help understand the development

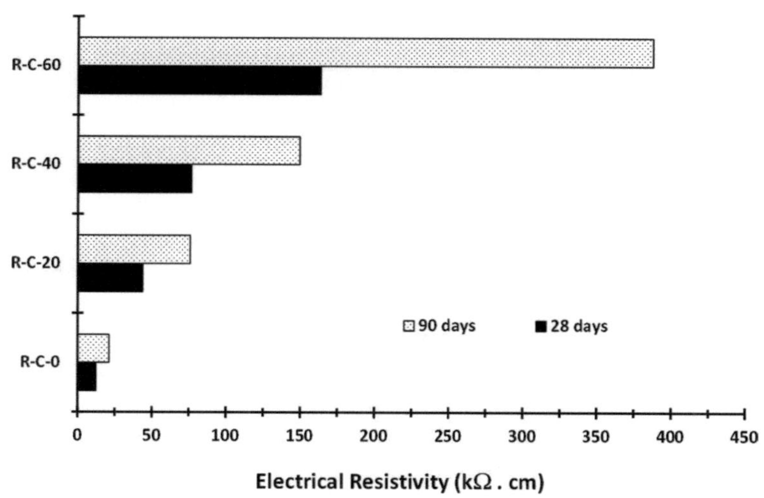

Figure 21.
Electrical resistivity of SCC.

Mixture	Porosity (%)	Median pore diameter* (μm)
R-C-0	24.989	8.1265
R-C-20	17.737	5.3136
R-C-40	15.604	3.9109
R-C-60	13.304	2.5002

Based on the intruded volume.

Table 8.
MIP results at 90 days of age (modified from [42]).

of the concrete microstructure and can also explain the other obtained results. **Table 8** shows the MIP test results at 90 days of age. Test results show that high CWP content has a significant reduction of the pore volume and the pores' size. The reduction in the pore volume and the pores' size indicates densification of the microstructure. Also, the MIP results confirm the improvement observed in the resistivity results and compressive strength.

5. Zero-cement alkali activated concrete (AAC)

Zero-cement alkali-activated concrete (AAC) emerged as an alternative to cement-based concrete [78–84]. Sometimes, AAC is referred to as inorganic-polymer or geopolymer concrete. In AAC, cement is completely replaced. AAC utilizes and silica and alumina rich materials to be alkali-activated to form a three-dimensional CaO-free alumino-silicate binder. AAC offers a significant opportunity for the reuse of several industrial by-products and wastes such as fly ash, metakaolin, and blast-furnace slag. Geopolymerization technology is based on the reaction of alkaline solutions such as sodium hydroxide (NaOH), potassium hydroxide (KOH) and sodium silicate solution. The CWP is characterized by its high silica and alumina content which makes it a good candidate to be used in making ACC. The limited studies on suing CWP in AAC [38–40] concluded that the optimum curing temperature ranges from 60 to 80°C, the curing period ranges between 24 and 48 hours, and the molarity of the alkali solution is 12 M.

The use of CWP in the making AAC still needs further investigations to develop a better understanding of its performance. CWP is used to make AAC using different alkali solutions, mainly NaOH and KOH. Several parameters are investigated which include alkaline solutions with 12 M concentration (i.e., NaOH alone, KOH alone and combination), CWP to aggregate ratio (i.e., 1:1.5–1:2.0–1:2.5), admixture dosage (i.e., 1.5 and 4.0%), curing time (i.e., 60°C for 24 and 48 hours), the inclusion of slag in addition to CWP (i.e., slag content 10, 20 and 40%). Several tests are used to evaluate the performance of the mixtures which include flowability (i.e., ASTM C1437 [85]), cube compressive strength, permeable pores (i.e., ASTM C642 [54]), initial rate of water absorption (i.e., ASTM C1585 [86]), and electrical resistivity (i.e., ASTM C1760 [53]). The COV ranged from 0.3 to 2.8%.

The sodium hydroxide flakes and potassium hydroxide are dissolved in distilled water to make a solution with the desired concentration (i.e., 12 M) at least 1 day before its use. **Table 9** shows the alkali solutions used and the combination of NaOH and KOH solutions. The dry ingredients are first mixed for about 1 minute. The sodium hydroxide and potassium hydroxide solutions are added to the dry materials based on the order of mixing in **Table 9** and mixed for 3 minutes.

I.D.	Alkali solutions %		Mixing regime of the solutions with the CWP
	KOH	NaOH	
A	0	100	—
B	100	0	—
C	20	80	NaOH solution is added first and mixed with solids for 1 minute, then KOH is added and mixing continues for an additional 2 minutes
D	40	60	NaOH and KOH solutions are mixed then added to solids and mixed for 3 minutes
E	60	40	KOH solution is added first and mixed with solids for 1 minute, then NaOH is added and mixing continues for an additional 2 minutes

Table 9.
Mixtures' I.D., alkali solutions used and mixing regime of solutions.

5.1 Effect of aggregate content

The effect of aggregate content was evaluated by the flowability and 7 days compressive strength. Mixtures are cured at 60°C for 24 hours. **Figure 22** shows the flowability and 7 days compressive strength as affected by the CWP to aggregate ratio. It is noticed that the flowability decreases as the aggregate content increases. This is similar to the behavior cement concrete as the CWP content acts as a lubricant between aggregate particles. Oppositely the 7 days compressive strength improved by the increase of the aggregate content. The mixing regime of the solution affects the flowability and strength. The mixing regime (A) shows the best flowability performance while the other mixing regimes show similar flowability values. The mixing regimes (D) and (E) produce the highest compressive.

5.2 Effect of admixture content

Superplasticizer (i.e., polycarboxylic ether based) is added with a dosage of 1.5 and 4.0% of the CWP weight. The AAC mixture with CWP to the aggregate ratio (1:2.5) and 24 hours curing at 60°C is used to examine the effect of admixture dosage. Flowability and the 7 days compressive strength results are presented in **Table 10**. The use of 1.5% by weight superplasticizer, shows variable improvement in the flowability and marginal improvement in the strength. By increasing the admixture dosage to 4.0%, the flowability and strength are improved. For both admixture dosages, the mixing regimes (D) and (E) show the best flowability improvement and highest compressive strength.

5.3 Effect of curing time

The AAC mixture with CWP to aggregate ratio (1:2.5) and 4% admixture is used to examine the effect of curing time (i.e., 24 and 48 hours) at 60°C. **Figure 23** shows the effect of curing time on the 7 days compressive strength. The compressive strength increases as the curing time increases. A similar trend is reported for metakaolin-based AAC [87]. Although increasing the curing time improves the compressive strength, the application of shorter curing time is considered from the point of reducing the energy consumption.

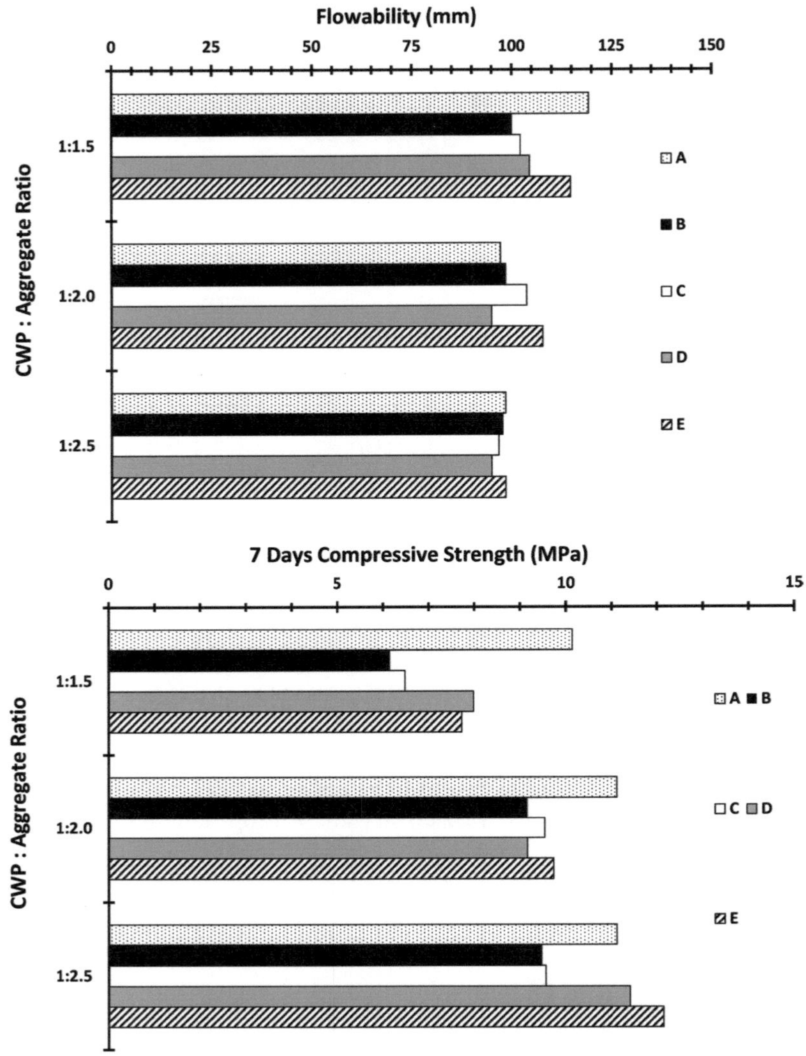

Figure 22.
Flowability and 7 days compressive strength as affected by CWP to aggregate ratio.

Mixing regime	Flowability (mm)		7 Days compressive strength (MPa)	
	1.5%*	4.0%*	1.5%*	4.0%*
A	95	107	12	14
B	95	107	10	12
C	99	110	11	13
D	110	116	13	15
E	112	117	14	16

Superplasticizer admixture dosage by weight of the CWP.

Table 10.
Effect of admixture on flowability and 7 days compressive strength.

5.4 Effect of slag content and curing regime

Several studies investigated the use of slag in making AAC [88–92]. Slag proved to be a suitable material in making AAC. Slag is characterized by having some

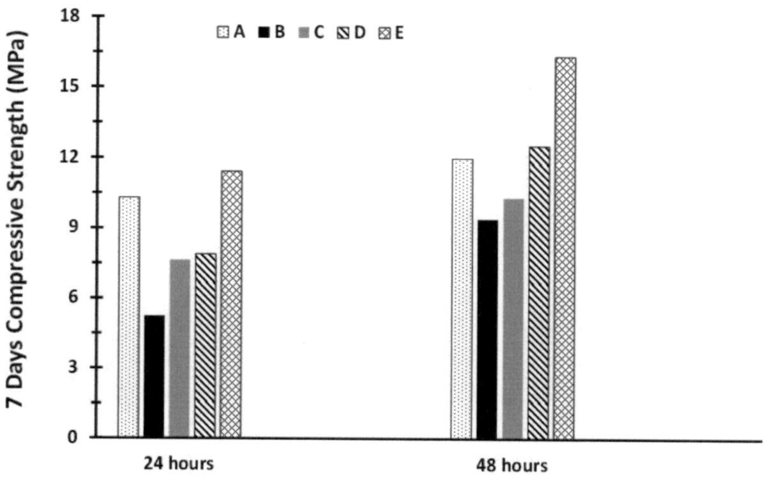

Figure 23.
Seven days compressive strength for the AAC mixture with CWP to aggregate ratio 1:2.5 as affected by curing time at 60°C.

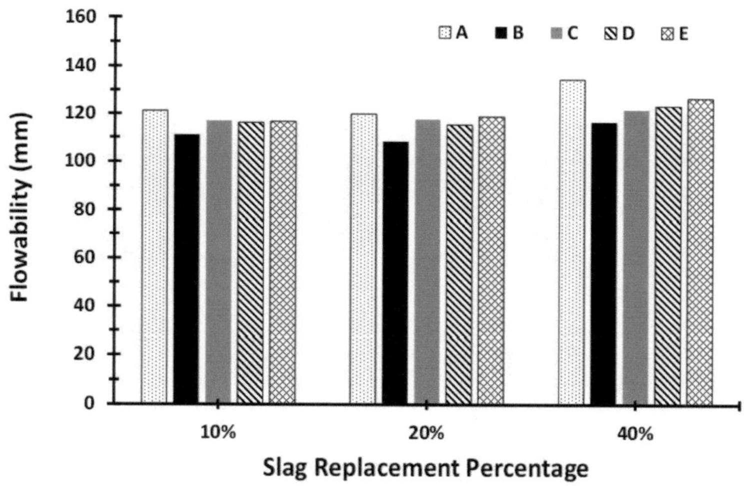

Figure 24.
Flowability of AAC including CWP and slag.

hydraulic reaction due to the existence of calcium oxide (CaO) beside the existence of silica and alumina for the alkali activation. Therefore, slag is used to replace part of the CWP. This will help improve the flowability of the AAC mixture and improve the strength development without the need to increase curing time. The AAC mixture with CWP to aggregate ration 1:2.5 and 4% admixture is used to assess the effect of including slag as a binder material in addition to the CWP. The slag replaced the CWP with 10, 20 and 40% by weight. The AAC mixtures including slag are subjected to three curing regimes; air curing, 24 hours at 60°C followed by air curing, and 24 hours at 60°C followed by water curing for 6 days. **Figure 24** shows the flowability of AAC mixtures including slag and CWP. The inclusion of slag improves the mixtures' flowability. The improvement is proportional to the slag content with the highest improvement at 40% slag.

The effect of including slag with CWP on the 7 days strength is displayed in **Figure 25**. The air cured mixtures showed the lowest strength development. It is observed that the (oven + air) and (oven + water) results are comparable for both the 20 and 40% slag replacements. The strength values are found to increase with the increase in slag % replacing the CWP, with the highest at 40% slag.

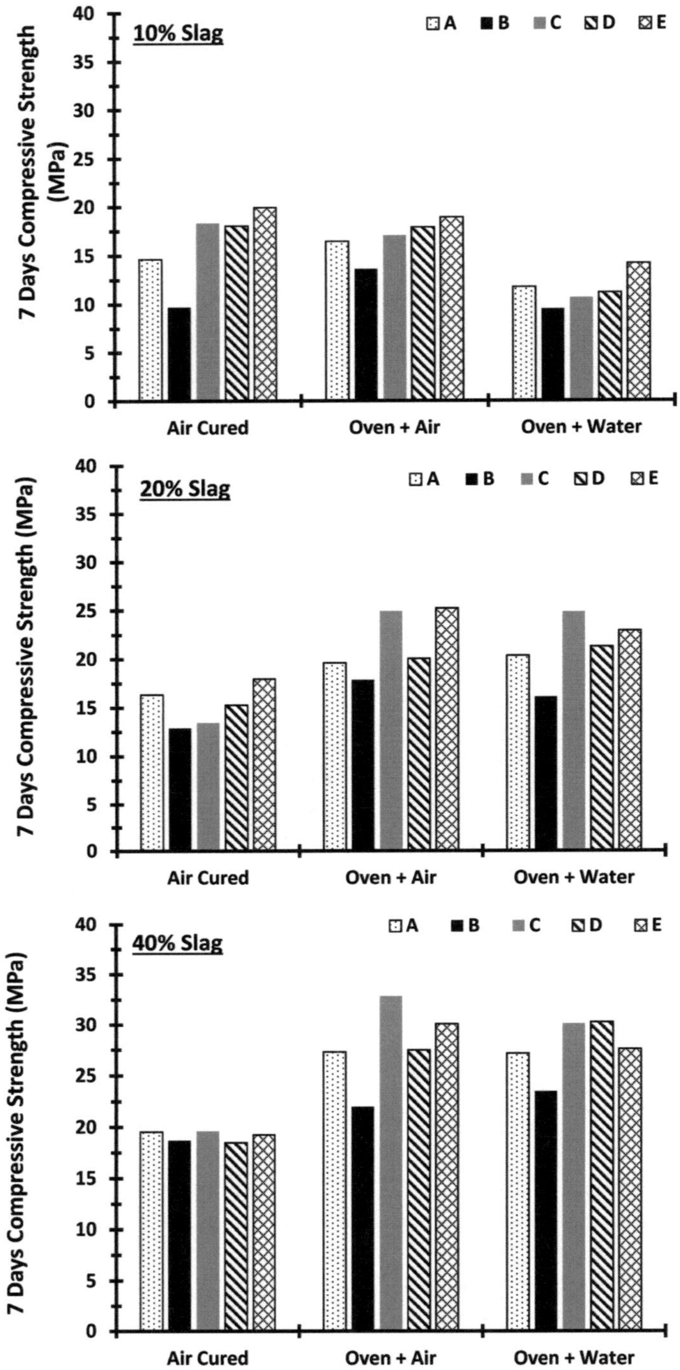

Figure 25.
Seven days compressive strength of AAC including CWP and slag.

The inclusion of slag is beneficial in producing AAC using CWP with a level of replacement of 40%. Based on the flowability and the 7 days compressive strength, the following are the optimum mixture's parameter to make AAC using CWP:

i. the CWP to the aggregate ratio is 2.5,

ii. the alkali solutions mixing regime (D) (i.e., NaOH 60% and KOH 40% mixed) produces suitable flowability and strength;

	Test age (days)	
	7	28
Compressive strength (MPa)	39.3	40.7
Permeable pores %	8.89	8.32
Electrical bulk resistivity (kΩ.cm)	17.9	18.2
Initial rate of absorption (mm/min$^{1/2}$) sorptivity	0.15	0.12

Table 11.
Seven and twenty-eight days results for optimum AAC mixture.

iii. the use of 4% of superplasticizer to improve flowability;

iv. the application of 24 hours at 60°C followed by air curing; and

v. the use of 40% by weight slag to replace CWP.

The performance of an AAC mixture following the above parameters is assessed. **Table 11** summarizes the obtained results. Results show that CWP in combination with 40% slag can produce AAC with strength suitable for different structural applications. The electrical resistivity and initial rate of absorption indicate that the produced AAC is characterized by high durability. The change in the test results values with age indicates that most of the reactions are finished at 7 days of age. Hence there is no need for waiting to evaluate the performance at 28 days of age similar to Portland cement concrete.

6. Conclusions

The CWP contains high silica and alumina content (i.e., >80%). Also, it is characterized by having some amorphous content which shows pozzolanic activity especially at late ages. Therefore, CWP has strong potentials to be used as an ingredient in making eco-friendly concretes.

Using CWP as an ingredient in making CVC is viable. High-performance concrete can be produced by including CWP as partial cement replacement. CWP improves the workability retention of the CVC mixtures. The inclusion of CWP will reduce the early-age strength and slowed the strength development. Significant improvement of CVC durability can be achieved by including high content of CWP. The CVC performance varies according to the CWP content. CWP can be used in the range of 10–20% to improve workability retention and late strength development. A CWP content ranging from 30 to 40% is needed to improve durability. If the performance of mixture requires the combination of workability retention, strength and durability, a CWP content ranging from 20 to 30% can be used to optimize all required characteristics.

CWP can be used as a partial cement replacement to produce SCC that meets international requirements. All fresh concrete properties, except for slump flow, are significantly improved by the incorporation of CWP. The improvement is proportional to the CWP content. Similar to CVC, the inclusion of CWP affected the strength development and enhanced the durability. SCC with improved fresh performance and optimized strength can be produced using 40% CWP as partial cement replacement.

The use of CWP in making AAC showed promising potentials. The production of AAC using CWP should consider the aggregate content of the mixture, the use of superplasticizer admixtures and the use of an alkali solution composed of NaOH and KOH. The combination of slag with CWP improves the workability and strength development without the need for long curing time to conserve energy. The combination of CWP with fly ash can also be an alternative to enhance the performance of the produced AAC.

Finally, CWP has encouraging potentials to be used as an ingredient to make eco-friendly conventional-vibrated concrete (CVC), self-compacting concrete (SCC) and zero-cement alkali-activated concrete (AAC). The concrete industry can and will play a vital role in the sustainable development through the utilization of industrial waste materials.

Acknowledgements

This work was financially supported by the UAEU-UPAR2 Research Grant # 31 N2018. Also, the donation of the ceramic waste powder for the study by PORCELLAN (ICAD II MUSSAFAH—ABU DHABI, UAE) is much valued. The help of master students Dima M. Kanaan and Sama T. Aly is highly appreciated. Support to the second author by Southern Plains Transportation Centre (SPTC) to University of New Mexico is much appreciated.

Author details

Amr S. El-Dieb[1]*, Mahmoud R. Taha[2] and Samir I. Abu-Eishah[3]

1 Civil and Environmental Engineering Department, United Arab Emirates University, Al Ain, UAE

2 Civil Engineering Department, University of New Mexico, Albuquerque, USA

3 Chemical and Petroleum Engineering Department, United Arab Emirates University, Al Ain, UAE

*Address all correspondence to: amr.eldieb@uaeu.ac.ae

References

[1] Chindaprasirt P, Chareerat T, Sirivivatnanon V. Workability and strength of coarse high calcium fly ash geopolymer. Cement and Concrete Composites. 2007;**29**(3):224-229. DOI: 10.1016/j.cemconcomp.2006.11.002

[2] Oacheco-Torgal F, Abdollahnejad Z, Camões AF, Jamshidi M, Ding Y. Durability of alkali-activated binders: A clear advantage over Portland cement or an unproven issue? Construction and Building Materials. 2012;**30**:400-405. DOI: 10.1016/j.conbuildmat.2011.12.017

[3] Van Deventer JSJ, Provis J, Duxson P. Technical and commercial progress in the adoption of geopolymer cement. Minerals Engineering. 2012;**29**:89-104. DOI: 10.1016/j.mineng.2011.09.009

[4] Batayneh M, Marie I, Asi I. Use of selected waste materials in concrete mixes. Waste Management. 2007;**27**(12):1870-1876. DOI: 10.1016/j.wasman.2006.07.026

[5] ACIMAC/MECS. World Production and Consumption of Ceramic Tiles [Internet]. 4th ed 2016. Available from: http://www.mec-studies.com/filealbum/583_0.pdf [Accessed: Oct 29, 2017]

[6] Cheng YH, Huang F, Liu R, Hou JL, Li GL. Test research on effects of waste ceramic polishing powder on the permeability resistance of concrete. Materials and Structures. 2016;**49**(3):729-738. DOI: 10.1617/s11527-015-0533-6

[7] De Brito J, Pereira AS, Correia JR. Mechanical behavior of non-structural concrete made with recycled ceramic aggregates. Cement and Concrete Composites. 2005;**27**(4):429-433. DOI: 10.1016/j.cemconcomp.2004.07.005

[8] Senthamarai RM, Manoharan PD. Concrete with ceramic waste aggregate. Cement and Concrete Composites. 2005;**27**(9):910-913. DOI: 10.1016/j.cemconcomp.2005.04.003

[9] Correia JR, De Brito J, Pereira AS. Effects on concrete durability of using recycled ceramic aggregates. Materials and Structures. 2006;**39**(2):169-177. DOI: 10.1617/s11527-005-9014-7

[10] Guerra I, Vivar I, Llamas B, Juan A, Moran J. Eco-efficient concretes: The effects of using recycled ceramic material from sanitary installations on the mechanical properties of concrete. Waste Management. 2009;**29**(2):643-646. DOI: 10.1016/j.wasman.2008.06.018

[11] Senthamarai RM, Manoharan PD, Gobinath D. Concrete made from ceramic industry waste: Durability properties. Construction and Building Materials. 2011;**25**(5):2413-2419. DOI: 10.1016/j.conbuildmat.2010.11.049

[12] Medina C, de Rojas MIS, Frías M. Reuse of sanitary ceramic wastes as coarse aggregate in eco-efficient concretes. Cement and Concrete Composites. 2012;**34**(1):48-54. DOI: 10.1016/j.cemconcomp.2011.08.015

[13] Medina C, Frías M, de Rojas MIS. Microstructure and properties of recycled concretes using ceramic sanitary ware industry waste as coarse aggregate. Construction and Building Materials. 2012;**31**:112-118. DOI: 10.1016/j.conbuildmat.2011.12.075

[14] Medina C, Frías M, de Rojas MIS, Thomas C, Polanco JA. Gas permeability in concrete containing recycled ceramic sanitary ware aggregate. Construction and Building Materials. 2012;**37**:597-605. DOI: 10.1016/j.conbuildmat.2012.08.023

[15] Medina C, de Rojas MIS, Frías M. Properties of recycled ceramic aggregate concretes: Water resistance. Cement and Concrete Composites. 2013;**40**:21-29. DOI: 10.1016/j.cemconcomp.2013.04.005

[16] Halicka A, Ogrodnik P, Zegardlo B. Using ceramic sanitary ware waste as concrete aggregate. Construction and Building Materials. 2013;**48**:295-305. DOI: 10.1016/j.conbuildmat.2013.06.063

[17] Lopez V, Llamas B, Juan A, Moran JM, Guerra I. Eco-efficient concretes: Impact of the use of white ceramic powder on the mechanical properties of concrete. Biosystems Engineering. 2007;**96**(4):559-564. DOI: 10.1016/j.biosystemseng.2007.01.004

[18] Higashiyama H, Sappakittipakorn M, Sano M, Yagishita F. Chloride ion penetration into mortar containing ceramic waste aggregate. Construction and Building Materials. 2012;**33**:48-54. DOI: 10.1016/j.conbuildmat.2012.01.018

[19] Higashiyama H, Yagishita F, Sano M, Takahashi O. Compressive strength and resistance to chloride penetration of mortars using ceramic waste as fine aggregate. Construction and Building Materials. 2012;**26**(1):96-101. DOI: 10.1016/j.conbuildmat.2011.05.008

[20] Higashiyama H, Yamauchi K, Sappakittipakorn M, Sano M, Takahashi O. A visual investigation on chloride ingress into ceramic waste aggregate mortars having different water to cement ratios. Construction and Building Materials. 2013;**40**:1021-1028. DOI: 10.1016/j.conbuildmat.2012.11.078

[21] Bahoria BV, Parbat DK, Naganaik PB. Replacement of natural sand in concrete by waste products: A state of art. Journal of Environmental Research and Development. 2013;7(4A):1651-1656

[22] Jiménez JR, Ayuso J, López M, Fernández JM, de Brito J. Use of fine recycled aggregates from ceramic waste in masonry mortar manufacturing. Construction and Building Materials. 2013;**40**:679-690. DOI: 10.1016/j.conbuildmat.2012.11.036

[23] Vejmelková E, Keppert M, Rovnaníková P, Ondráček M, Keršner Z, Černý R. Properties of high performance concrete containing fine-ground ceramics as supplementary cementitious material. Cement and Concrete Composites. 2012;**34**(1):55-61. DOI: 10.1016/j.cemconcomp.2011.09.018

[24] Rahhal V, Irassar E, Castellano C, Pavslík Z, Černý R. Utilization of ceramic wastes as replacement of Portland cements. In: Proceedings of the International Conference on Construction Materials and Structures (ICCMATS-1); 24-26 November 2014; South Africa. Amsterdam: IOS Press; 2014. pp. 208-213

[25] Steiner LR, Bernardin AM, Pelisser F. Effectiveness of ceramic tile polishing residues as supplementary cementitious materials for cement mortars. Sustainable Materials and Technologies. 2015;**4**:30-35. DOI: 10.1016/j.susmat.2015.05.001

[26] Irassar E, Rahhal V, Tironi A, Trezza M, Pavlík Z, Pavlíková M, et al. Utilization of ceramic wastes as pozzolanic materials. In: Technical Proceedings of the 2014 NSTI Nanotechnology Conference and Expo. Vol. 1-3. 15-18 June 2014. Washington DC: CRC Press; 2014. pp. 316-319

[27] Pokorný J, Fořt J, Pavlíková M, Studnika J, Pavlík Z. Application of mixed ceramic powder in cement based composites. Advanced Materials Research. 2014;**1054**:177-181. DOI: 10.4028/www.scientific.net/AMR.1054.177

[28] Vejmelková E, Koáková D, Kulovaná T, Hubáek A, Erný R. Mechanical and thermal properties

of moderate-strength concrete with ceramic powder used as supplementary cementitious material. Advanced Materials Research. 2014;**1054**:194-198. DOI: 10.4028/www.scientific.net/AMR.1054.194

[29] Heidari A, Tavakoli D. A study of the mechanical properties of ground ceramic powder concrete incorporating nano-SiO_2 particles. Construction and Building Materials. 2013;**38**:255-264. DOI: 10.1016/j.conbuildmat.2012.07.110

[30] Pacheco-Torgal F, Jalali S. Compressive strength and durability properties of ceramic wastes based concrete. Materials and Structures/Materiaux et Constructions. 2011;**44**:155-167. DOI: 10.1617/s11527-010-9616-6

[31] Reiterman P, Holcapek O, Cachova M, Vogel F, Jogl M, Konvalinka P. Basic and hygric properties of concrete containing fine ceramic powder. Advanced Materials Research. 2014;**897**:188-191. DOI: 10.4028/www.scientific.net/AMR.897.188

[32] Mishra N, Vasugi V. Experimental assessment of properties of ternary blended concrete using GGBS and ceramic powder. Indian Concrete Journal. 2015;**89**:74-80

[33] Cheng Y, Huang F, Li G-L, Xu L, Hou J-L. Test research on effects of ceramic polishing powder on carbonation and sulphate-corrosion resistance of concrete. Construction and Building Materials. 2014;**55**:440-446. DOI: 10.1016/j.conbuildmat.2014.01.023

[34] Bignozzi MC, Bonduá S. Alternative blended cement with ceramic residues: Corrosion resistance investigation on reinforced mortar. Cement and Concrete Research. 2011;**41**(9):647-654. DOI: 10.1016/j.cemconres.2011.05.001

[35] Fatima E, Jhamb A, Kumar R. Ceramic dust as construction material in rigid pavement. American Journal of Civil Engineering and Architecture. 2013;**1**(5):112-116. DOI: 10.12691/ajcea-1-5-5

[36] Subaşi S, Öztürk H, Emiroğlu M. Utilizing of waste ceramic powders as filler material in self-consolidating concrete. Construction and Building Materials. 2017;**149**:567-574. DOI: 10.1016/j.conbuildmat.2017.05.180

[37] Jerônimo VL, Meira GR, LCPd-S F. Performance of self-compacting concretes with wastes from heavy ceramic industry against corrosion by chlorides. Construction and Building Materials. 2018;**169**:900-910. DOI: 10.1016/j.conbuildmat.2018.03.034

[38] Sun Z, Cui H, An H, Tao D, Xu Y, Zhai J, et al. Synthesis and thermal behavior of geopolymer-type material from waste ceramic. Construction and Building Materials. 2013;**49**:281-287. DOI: 10.1016/j.conbuildmat.2013.08.063

[39] El-Dieb AS, Shehab IE. Cementless concrete using ceramic waste powder. In: Proceedings of the International Conference on Construction Materials and Structures (ICCMATS-1); 24-26 November 2014; South Africa. Amsterdam: IOS Press; 2014. pp. 487-494

[40] Aly ST, Kanaan DM, El-Dieb AS, Abu-Eishah SI. Properties of ceramic waste powder based geopolymer concrete. In: Proceedings of the International Congress on Polymers in Concrete (ICPIC 2018); 29 April–1 May 2018. Washington DC: Springer; 2018. pp. 429-436

[41] Aly ST, El-Dieb AS, Reda Taha MM. Self-compacting concrete with large ceramic waste powder content as a partial replacement of cement. Journal of Materials in Civil Engineering. 2018 (accepted for publication in ASCE)

[42] Aly ST, El-Dieb AS, Reda Taha MM. Ceramic waste powder for eco-friendly self-compacting concrete. ASTM Advances in Civil Engineering Materials. 2018;**7**(1):426-445. DOI: 10.1520/ACEM20180043

[43] El-Dieb AS, Kanaan DM. Ceramic waste powder an alternative cement replacement—Characterization and evaluation. Journal of Sustainable Materials and Technology. 2018;**17**:e00063. DOI: 10.1016/j.susmat.2018.e00063

[44] El-Dieb AS, Reda Taha MM, Kanaan DM, Aly ST. Ceramic waste powder from landfill to sustainable concretes. ICE Construction Materials. 2018;**171**(3):109-116. DOI: 10.1680/jcoma.17.00019

[45] Kannan DM, Aboubakr SH, El-Dieb AS, Reda Taha MM. High performance concrete incorporating ceramic waste powder as large partial replacement of Portland cement. Construction and Building Materials. 2017;**144**:35-41. DOI: 10.1016/j.conbuildmat.2017.03.115

[46] ASTM C618-12a. Standard Specification for Coal Fly Ash and Raw or Calcined Natural Pozzolan for Use in Concrete. West Conshohocken, Pennsylvania: ASTM International; 2017

[47] ASTM C311/C311M-13. Standard Test Methods for Sampling and Testing Fly Ash or Natural Pozzolans for Use in Portland-Cement Concrete, ASTM International. West Conshohocken, Pennsylvania: ASTM International; 2017

[48] Chopra D, Siddique R. Strength, permeability and microstructure of self-compacting concrete containing rice husk ash. Biosystems Engineering. 2015;**130**:72-80. DOI: 10.1016/j.biosystemseng.2014.12.005

[49] Donatello S, Tyrer M, Cheeseman CR. Comparison of test methods to assess pozzolanic activity. Cement and Concrete Composites. 2010;**32**(2):121-127. DOI: 10.1016/j.cemconcomp.2009.10.008

[50] BS EN 196-5:2011. Methods of Testing Cement—Part 5: Pozzolanicity Test for Pozzolanic Cement. London, UK: British Standards Institution; 2011

[51] ASTM C143/C143M-15a. Standard Test Method for Slump of Hydraulic-Cement Concrete. West Conshohocken, Pennsylvania: ASTM International; 2017

[52] ASTM C1202-17. Standard Test Method for Electrical Indication of Concrete's Ability to Resist Chloride Ion Penetration. West Conshohocken, Pennsylvania: ASTM International; 2017

[53] ASTM C1760-12. Standard Test Method for Bulk Electrical Conductivity of Hardened Concrete. West Conshohocken, Pennsylvania: ASTM International; 2017

[54] ASTM C642-13. Standard Test Method for Density, Absorption, and Voids in Hardened Concrete. West Conshohocken, Pennsylvania: ASTM International; 2017

[55] Mehta PK. Concrete Structure, Properties and Materials. New Jersey, USA: Prentice-Hall, Inc.; 1986. p. 450

[56] Bektas F. Use of ground clay brick as a supplementary cementitious material in concrete-hydration characteristics, mechanical properties, and ASR durability [dissertation]. Civil Construction and Environmental Engineering: Iowa State University, Ames, Iowa 50011, USA; 2007

[57] ACI 222R-01. Protection of Metals in Concrete against Corrosion. Farmington Hills, USA: American Concrete Institute; 2009

[58] Shahroodi A. Development of test methods for assessment of concrete

durability for use in performance-based specifications [dissertation]. Department of Civil Engineering: University of Toronto; Toronto, Ontario M5S 1A4, Canada; 2010

[59] Zhao H, Sun W, Wu X, Gao B. The properties of the self-compacting concrete with fly ash and ground blast furnace slag mineral admixtures. Journal of Cleaner Production. 2015;**95**:66-74. DOI: 10.1016/j.jclepro.2015.02.050

[60] ACI 237R-07. Self-Consolidating Concrete. Fragminton Hills, MI: American Concrete Institute; 2009

[61] Sideris KK, Manita P. Mechanical characteristics and durability of self-consolidating concretes produced with no additional fine materials. Journal of Sustainable Cement-Based Materials. 2014;**3**(3-4):234-244. DOI: 10.1080/21650373.2014.924040

[62] García-Taengua E, Sonebi M, Crossett P, Taylor S, Deegan P, Ferrara L, et al. Performance of sustainable SCC mixes with mineral additions for use in precast concrete industry. Journal of Sustainable Cement-Based Materials. 2016;**5**(3):157-175. DOI: 10.1080/21650373.2015.1024297

[63] Zhu W, Gibbs JC. Use of different limestone and chalk powders in self-compacting concrete. Cement and Concrete Research. 2005;**35**(8):1457-1462. DOI: 10.1016/j.cemconres.2004.07.001

[64] Gesoğlu M, Güneyisi E, Erdoğan Ö. Properties of self-compacting concrete made with binary, ternary, and quaternary cementitious blends of fly ash, blast furnace slag, and silica fume. Construction and Building Materials. 2009;**23**(5):1847-1854. DOI: 10.1016/j.conbuildmat.2008.09.015

[65] Topcu IB, Bilir B, Uygunoğlu T. Effect of waste marble dust content as filler on properties of self-compacting concrete. Construction and Building Materials. 2009;**23**(5):1947-1953. DOI: 10.1016/j.conbuildmat.2008.09.007

[66] Liu M. Self-compacting concrete with different levels of pulverized fuel ash. Construction and Building Materials. 2010;**24**(7):1245-1252. DOI: 10.1016/j.conbuildmat.2009.12.012

[67] Liu M. Incorporating ground glass in self-compacting concrete. Construction and Building Materials. 2011;**25**(2):919-925. DOI: 10.1016/j.conbuildmat.2010.06.092

[68] Uysal M, Sumer M. Performance of self-compacting concrete containing different mineral admixtures. Construction and Building Materials. 2011;**25**(11):4112-4120. DOI: 10.1016/j.conbuildmat.2011.04.032

[69] Mandanoust R, Mousavi S. Fresh and hardened properties of self-compacting concrete containing metakaolin. Construction and Building Materials. 2012;**35**:752-760. DOI: 10.1016/j.conbuildmat.2012.04.109

[70] Beycioğlu A, Aruntaş HY. Workability and mechanical properties of self-compacting concretes containing LLFA, GBFS and MC. Construction and Building Materials. 2014;**73**:626-635. DOI: 10.1016/j.conbuildmat.2014.09.071

[71] EFNARC. The European Guidelines for Self-Compacting Concrete: Specification, Production and Use [Internet]. 2005. Available from: http://www.efnarc.org/pdf/SCCGuidelinesMay2005 [Accessed: Oct 1, 2017]

[72] ASTM C1611/C1611M-14. Standard Test Method for Slump Flow of Self-Consolidating Concrete. West Conshohocken, Pennsylvania: ASTM International; 2017

[73] ASTM C1621/C1621M-17. Standard Test Method for Passing Ability of

Self-Consolidating Concrete by J-Ring. West Conshohocken, Pennsylvania: ASTM International; 2017

[74] ASTM C1610/C1610M-17. Standard Test Method for Static Segregation of Self-Consolidating Concrete Using Column Technique. West Conshohocken, Pennsylvania: ASTM International; 2017

[75] Sfikas IP, Badogiannis EG, Trezos KG. Rheology and mechanical characteristics of self-compacting concrete mixtures containing metakaolin. Construction and Building Materials. 2014;**64**:121-129. DOI: 10.1016/j.conbuildmat.2014.04.048

[76] Le HT, Ludwig H-M. Effect of rice husk ash and other mineral admixtures on properties of self-compacting high performance concrete. Materials & Design. 2016;**89**:156-166. DOI: 10.1016/j.matdes.2015.09.120

[77] Mehta PK, Monteiro PJM. Concrete: Microstructure, Properties, and Materials. 3rd ed. McGraw Hill; 2006. p. 659

[78] Shehab HK, Eisa AS, Wahba AM. Mechanical properties of fly ash based geopolymer concrete with full and partial cement replacement. Construction and Building Materials. 2016;**126**:560-565. DOI: 10.1016/j.conbuildmat.2016.09.059

[79] Mehta A, Siddique R. Properties of low-calcium fly ash based geopolymer concrete incorporating OPC as partial replacement of fly ash. Construction and Building Materials. 2017;**150**:792-807. DOI: 10.1016/j.conbuildmat.2017.06.067

[80] Pacheco-Torgal F, Castro-Gomes J, Jalali S. Alkali-activated binders: A review. Part 1. Historical background, terminology, reaction mechanisms and hydration products. Construction and Building Materials. 2008;**22**:1305-1314. DOI: 10.1016/j.conbuildmat.2007.10.015

[81] Pacheco-Torgal F, Castro-Gomes J, Jalali S. Alkali-activated binders: A review. Part 2. About materials and binders manufacture. Construction and Building Materials. 2008;**22**:1315-1322. DOI: 10.1016/j.conbuildmat.2007.03.019

[82] Shi C, Jiménez AF, Palomo A. New cements for the 21st century: The pursuit of an alternative to Portland cement. Cement and Concrete Research. 2011;**41**:750-761. DOI: 10.1016/j.cemconres.2011.03.016

[83] Duxson P, Provis JL, Lukey GC, Van Deventer JSJ. The role of inorganic technology in the development of 'Green Concrete'. Cement and Concrete Research. 2007;**37**(12):590-1597. DOI: 10.1016/j.cemconres.2007.08.018

[84] Komnitsas KA. Potential of geopolymer technology towards green buildings and sustainable cities. Procedia Engineering. 2011;**21**:1023-1032. DOI: 10.1016/j.proeng.2011.11.2108

[85] ASTM C1437-15. Standard Test Method for Flow of Hydraulic Cement Mortar. West Conshohocken, Pennsylvania: ASTM International; 2017

[86] ASTM C1585-13. Standard Test Method for Measurement of Rate of Absorption of Water by Hydraulic-Cement Concretes. West Conshohocken, Pennsylvania: ASTM International; 2017

[87] Ronaník P. Effect of curing temperature on the development of hard structure of metakaolin-based geopolymer. Construction and Building Materials. 2010;**24**(7):1176-1183. DOI: 10.1016/j.conbuildmat.2009.12.023

[88] Nath P, Sarker PK. Effect of GGBFS on setting, workability and early strength properties of fly ash

geopolymer concrete cured in ambient condition. Construction and Building Materials. 2014;**66**:163-171. DOI: 10.1016/j.conbuildmat.2014.05.080

[89] Puertas F, Fernandez-Jimenez A. Mineralogical and microstructural characterization of alkali-activated fly ash/slag pastes. Cement and Concrete Composites. 2003;**25**(3):287-292. DOI: 10.1016/S0958-9465(02)00059-8

[90] Puertas F, Martínez-Ramírez S, Alonso S, Vázquez T. Alkali-activated fly ash/slag cement strength behavior and hydration products. Cement and Concrete Research. 2000;**30**(10):1625-1632. DOI: 10.1016/S0008-8846(00)00298-2

[91] Akçaözoğlu S, Atiş CD. Effect of granulated blast furnace slag and fly ash addition on the strength properties of lightweight mortars containing waste PET aggregates. Construction and Building Materials. 2011;**25**(10):4052-4058. DOI: 10.1016/j.conbuildmat.2011.04.042

[92] Wang SD, Scrivener KL. ^{29}Si and ^{27}Al NMR study of alkali-activated slag. Cement and Concrete Research. 2003;**33**(5):769-774. DOI: 10.1016/S0008-8846(02)01044-X

Chapter 3

Synthesis of High-Purity Ceramic Nano-Powders Using Dissolution Method

Suminar Pratapa, Ella A.D. Kiswanti, Dien R. Diana, Yufi Hariyani, Lisma D.K. Sari, Musyarofah Musyarofah, Triwikantoro Triwikantoro and Malik A. Baqiya

Abstract

A set of ceramic powders has been synthesized using a "bottom-up" approach which is denoted here as the dissolution method. The raw materials were metal powders or minerals. The dissolution media were strong acid or base solutions. In the case of metallic raw materials, magnesium and titanium powders were separately dissolved in hydrochloric acid to obtain their precursors. They were then dried, washed, and calcined in air at various temperatures to produce pure MgO and TiO_2 nano-powders. Pure $MgTiO_3$ nano-powders by mixing the precursors at the stoichiometric ratio and calcining the dried mixture at a temperature as low as 700°C have also been successfully synthesized. In the mineral case, local zircon sand was used as the raw material. A standard procedure to extract the "clean" and pure zircon powder was applied which included washing, magnetic separation, and reactions using hydrochloric acid and sodium hydroxide. A pure zircon nano-powder was obtained by applying mechanical ball-milling to the zircon powder. The zircon powder was also chemically dissociated to give amorphous silica (SiO_2), cristobalite, amorphous zirconia (ZrO_2), and nanometric tetragonal zirconia powders.

Keywords: dissolution method, strong acid and base, metal and mineral, ceramic nano-powders

1. Introduction

Many efforts have been paid to produce nano-powders since they exhibit different and usually outstanding physical and chemical properties as compared to their larger counterparts. Ceramic nano-powders are even more attractive since they are thermally and chemically more stable than non-ceramic ones. Several applications of such nano-powders are in drug delivery, corrosion inhibitor, catalyst, and microwave communication. Oxide ceramics such as MgO, TiO_2, $MgTiO_3$, ZrO_2, and SiO_2 are more abundant than non-oxide ceramics. Their use in technological and industrial applications is hence more substantial.

Oxide ceramic powders are not easily found in nature in their simple mono- and bi-cationic forms. Yet, their technological benefits are very valuable. Their existence in nature is usually in the form of complex compounds and needs further processing to achieve high-purity substances. Furthermore, natural nano-ceramic powders are hardly found. As a result, various approaches have been proposed to synthesize such materials. Two general ways were usually used, that is, bottom-up and top-down methods. The former requires precursors of the desired cation(s) and usually uses heating in air or oxygen-controlled environment to develop the ceramics. The latter is basically a "breakdown" approach of a larger ceramic grains or particles by milling.

By definition, a nanometric powder means it has crystallite or grain or particle size less than 100 nm, although some researchers claimed that sub-nanometric size of <200 nm was still acceptable.

Several examples of synthesis of oxide nano-ceramic powders are solvothermal, sol-gel, and (co-)precipitation methods. The solvothermal involves the use of (usually) nonaqueous precursors and an autoclave to produce nanoparticles with unique microstructures. This method, for example, has been used to produce nanorods [1, 2], nanoclusters [3], and hollow spheres [4]. It is, however, a complex procedure. Meanwhile, the sol-gel method includes the use of complex precursors as the raw materials. For instance, synthesis of nanoparticles of magnesium and titanium oxides [5–8] has been reported recently. Despite its potential in controlling the size and shape of the products, the sol-gel process usually is time-consuming and costly. Finally, the precipitation or coprecipitation method has been reported by several researchers as an effective method to produce magnesia, titania (anatase), and zircon nano-powders [9–11]. The use of a precursor, washing with a certain liquid (usually distilled water), drying in air, and calcination are basic attributes in coprecipitation synthesis. The crystallite size of the synthesized powders for each method depends on many factors, particularly the type of precursors, as well as media, time, and temperature for processing. Some examples of nano-ceramics synthesized by these methods are presented in **Table 1**. Examples of bi-cationic ceramic nano-powders are also given.

Recently, an approach to processing oxide nano-ceramic powders has been developed. The approach, here designated as the dissolution method, comprises dissolution of a raw powder into a strong acid, followed by drying and washing, and finally calcination in air. There were several pure metals and minerals under our

Nano-material	Precursor	Method	Calcination temperature (°C)	Crystallite size (nm)	Refs.
MgO	$Mg(NO_3)_2.6H_2O$	Sol-gel	500	30	[5]
	$MgCl_2$	Sol-gel	800	100	[12]
	$Mg(CH_3COO)_2.4H_2O$	Solid	800	53	[13]
TiO_2	Tetrabutyl titanate	Sol-gel	160	6	[14]
	Titanium tetraisopropoxide	Sol-gel	500	30	[15]
	$TiOSO_4/CO(NH_2)_2$	Hydrothermal	1000	9	[10]
$MgTiO_3$	$Ti(OH)_4$ + Mg $(NO_3)_2.6H_2O$	Coprecipitation	700	30–40	[16]
	Mg $(NO_3)_2{\cdot}6H_2O$ + Ti $[OCH(CH_3)_2]_4$	Sol-gel	400	12	[17]
Mg_2SiO_4	$(CH_3COO)_2Mg.4H_2O$ and $C_8H_{20}O_4Si$	Sol-gel	900	33	[18]

Table 1.
Methods to produce mono- and bi-cationic oxide nano-ceramics from references.

study. In terms of synthesis of nano-materials, it is also classified as a "bottom-up" approach and a wet method.

The report of the synthesis is arranged by the type of the raw materials, that is, metal and mineral. The metal powders were magnesium and titanium, which produced magnesia, titania, and magnesium titanate nano-powders. The second group materials were synthesized with a slightly different way, that is, there were cleaning and demagnetization to improve the "purity" of the raw material. Further approaches were used to obtain nanometric zircon, silica, and zirconia powders. The success of the syntheses was affirmed by analytical methods like thermogravimetric and differential thermal analyses, X-ray fluorescence spectroscopy, X-ray diffractometry, and high-resolution transmission electron microscopy.

2. Dissolution method

In principle, dissolution of solids into a liquid or other solvents is a process by which the original states become dissolved components (solutes), hence forming a solution of the solid in the original solvent (see the schematic diagram in **Figure 1**). When a dissolution occurs, the dissolved component separates into ions or molecules, and each of them is surrounded by the molecules of the solvent. Using the dissolution process, one can generate a precursor of a cation from a metal or a mineral if it is soluble in a selected (strong) acid. For example, magnesium reacts with hydrochloric acid according to $Mg(s) + 2HCl(aq) \rightarrow MgCl_2\ (aq) + H_2(g)$ where the hydrogen gas is released [19]. On the other hand, titanium is a rather unreactive metal, making it difficult to dissolve unless more external energy such as heat is provided. Dissolving titanium powder in hydrochloric acid is possible as long as the process is run at approximately 60–70°C where the product is a purple solution of titanium trichloride.

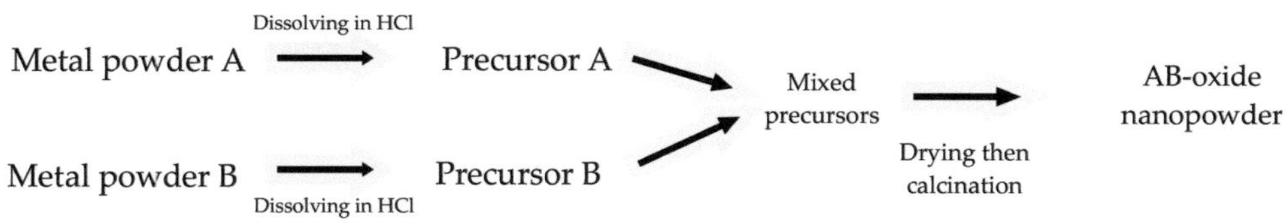

Figure 1.
Schematic diagram for dissolution process, an example for separate dissolution of A and B powders in HCl to synthesize AB oxide nano-powder.

Dissolution of metal oxide MO is also possible [20, 21], where M denotes a metal. Several factors may affect the dissolution kinetics including physical form and constitution of the oxides, as well as pH (acid or base), redox potential, chelating strength, concentration, and temperature of the solution. Examples of metal oxide dissolution in strong acids are for lanthanum oxide [22], iron oxides [23], and zinc ferrite [24]. In this work, the dissolution method was further used for selectively extracting the cations from natural mineral to produce high-purity ceramic nano-powders.

3. Synthesis and phase analyses of the powders

Tables 2 and **3** present a series of metal oxide powders which were produced by metal and mineral dissolutions, respectively. For the metal-dissolved powders, hydrochloric acid was used as the solvent. The oxide precursors were obtained by

Metal powder	Acid/ base	Precursor	Calcination temperature (°C)	Product	Crystallite size (nm)
Mg	HCl	$MgCl_2$	700	Pure MgO	98
Ti	HCl	$TiCl_4$	200 < T < 700	TiO_2 anatase	96–114
				TiO_2 rutile	6–96
			>800	TiO_2 rutile	>200
Ti / Mg	HCl	Mixture of $TiCl_4$ and $MgCl_2$	700	Pure $MgTiO_3$	65

Table 2.
Dissolution of metal powders in acid to produce nanocrystalline oxide powders.

drying the precipitates after subsequent washing with water. The solvent for processing zircon sand, as shown in **Table 3**, depends on the purposed powder. Zircon sand can be processed to obtain nano-zircon, silica, or nano-zirconia powders. The calcination temperatures were selected after inspecting the thermal data from the precursors.

The structure, composition, and nanocrystallinity of the powders were examined using CuKα X-ray diffraction (XRD) data to which the Rietveld-based *Rietica* [25] and Material Analysis Using Diffraction (*MAUD*) [26] softwares were employed. The former was applied for phase composition calculation according to the "ZMV" method, while the latter for average crystallite size estimation by implementing the Thomson-Cox-Hastings [14] model in a pseudo-Voigt peak-shape function after instrumental correction using an annealed Y_2O_3 as suggested previously.

The relative weight fraction of phase i at each depth was determined by:

$$W_i = \frac{s_i(ZMV)_i}{\sum_{j=1}^{n} s_j(ZMV)_j} \tag{1}$$

where s_i denotes the Rietveld scale factors of sample i and n is the number of the phases. Z_i is the number of formula unit of phase i (calculated from the refined lattice parameters) with mass M_i in the unit cell volume V_i.

Mineral	Acid/base	Precursor/ precipitate	Calcination temperature (°C)	Product	Crystallite size (nm)
$ZrSiO_4$ (zircon sand)	NaOH	Na_2SiO_3	100 (drying)	Am. SiO_2	—
	HCl	Na_2SiO_3	900	Cristobalite	>500
	Milling in ethanol	—	100 (drying)	$ZrSiO_4$	160[a] 40[b]
	NaOH + HCl	$ZrOCl_2$	700	ZrO_2	20

[a] *Unmilled powder.*
[b] *Milled powder for 10 h at 150 rpm.*

Table 3.
Dissolution of natural zircon mineral to produce high-purity oxide powders.

High-resolution transmission electron microscope (TEM) was used to investigate the crystallite and particle sizes as well as the morphology of the synthesized powders.

3.1 Magnesium oxide (MgO)

The dried powder of Mg-dissolved solution was run for DTA/TG characterization where the associated plot is given in **Figure 2**. A significant drop of the mass of the powdered sample between RT and approximately 600°C is related to some thermodynamic phenomena recorded as endothermic peaks. The peaks between RT and 300°C are attributed to release water molecules with a mass loss of approximately 50%, while the peak at ~500°C is due to the liberation of chloride ions and molecules. The last release is also indicated by some mass drop up to 600°C. The DTA/TG observations led the synthesis of MgO. A calcination temperature range of 400–800°C was then selected. The success of the syntheses was examined mainly using XRD measurements (**Figure 3**).

The XRD patterns of the calcined Mg-dissolved powders clearly show the phase evolution with calcination temperature. At 400°C, only bischofite ($MgCl_2{\cdot}6H_2O$) was detected. Increasing the temperature to 500°C resulted in the formation of MgO, and further calcination at 700°C gave pure MgO. This result indicates that between 400 and 700°C, there was a reaction of $MgCl_2{\cdot}6H_2O$ (s) $\rightarrow$ MgO (s) + H_2O (g) + Cl_2 (g). As implied by the DTA/TG data, there is a drop of mass and exothermic phenomena which can be ascribed to this reaction. Further analysis shows that the composition and crystallite size of the products change subsequently (**Table 4**). Therefore, the Mg metal-dissolved method can produce nanocrystalline, pure MgO powder after calcination at 700°C. Calcination at 800°C does not significantly alter the phase characteristics of the powder.

3.2 Titanium dioxide (TiO_2)

DTA/TG plot of the dried Ti-dissolved solution is given in **Figure 4**. The XRD patterns of the calcined powders are presented in **Figure 5**. Similar behavior in the phase formation of MgO was found here for the Ti-dissolved powders. Significant mass release occurs and followed by crystallite formation. Interestingly, the dried Ti-dissolved powder has already exhibited rutile with a relatively low degree of crystallinity as indicated by low diffraction intensities. The crystallinity becomes clearer at higher calcination temperatures. Pure rutile is achieved at 400°C, but anatase forms above 600°C. In most cases, the formation of anatase occurs at a

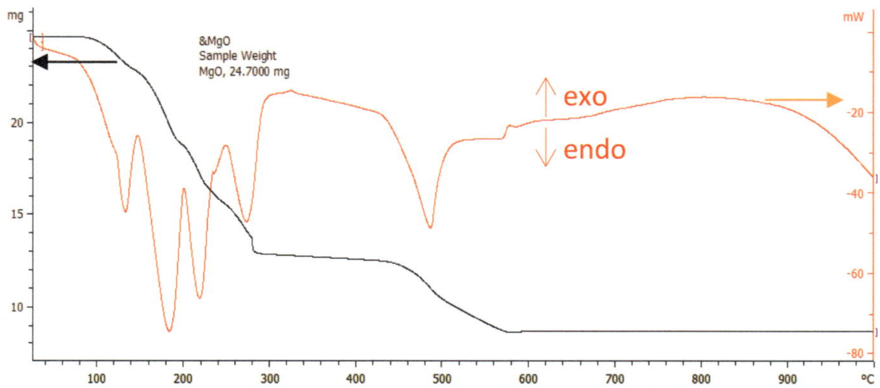

Figure 2.
DTA/TG plot of the Mg-dissolved dried powder.

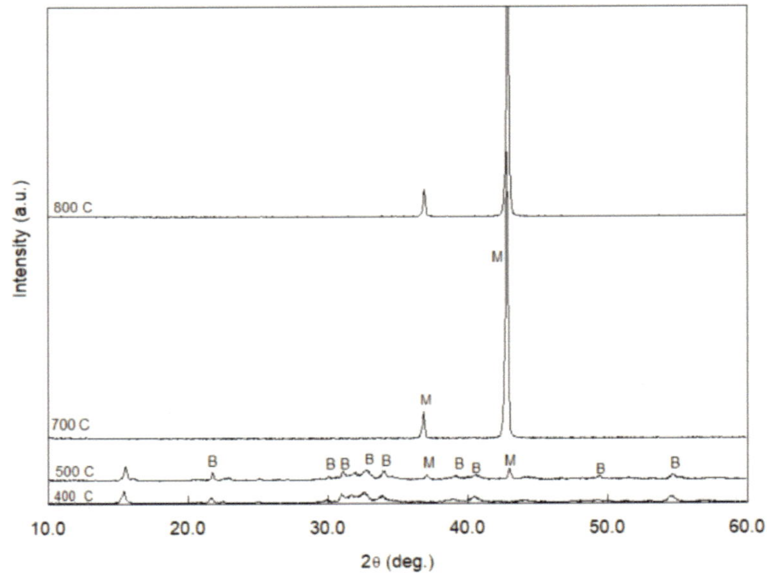

Figure 3.
X-ray diffraction patterns (CuKα radiation) of the calcined Mg-dissolved powders. Symbols: B = bischofite ($MgCl_2{\cdot}6H_2O$), M = magnesia (MgO, also known as periclase).

Temperature	Weight fraction (%)		Crystallite size (nm)	
	MgO (periclase)	$MgCl_2.6H_2O$ (bischofite)	MgO (periclase)	$MgCl_2.6H_2O$ (bischofite)
400°C	—	100	—	90(12)
500°C	18.5	81.5	99(5)	97(5)
700°C	100	—	98(5)	—
800°C	100	—	98(3)	—

Table 4.
Phase weight fraction and MgO crystallite size of the calcined Mg-dissolved powders extracted from XRD data.

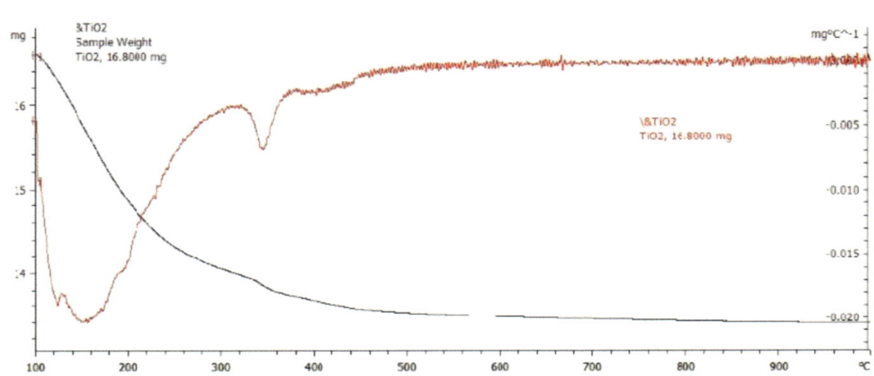

Figure 4.
DTA/TG plot of the Ti-dissolved dried powder.

lower temperature than rutile [8, 10]. In our study, the formation of rutile is more spontaneous than anatase possibly due to the type of precursor and its environment. A similar result was observed by others [27]. It is also worth noting that the broad

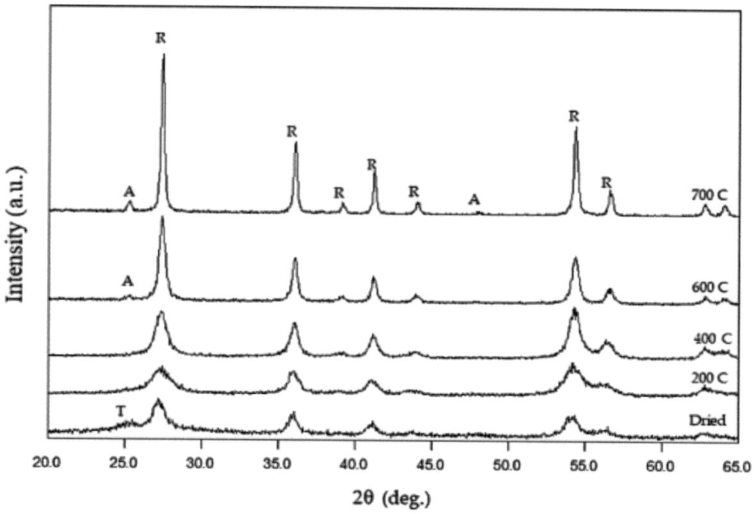

Figure 5.
X-ray diffraction patterns (CuKα radiation) of the calcined Ti-dissolved powders. Symbols: A = TiO_2 anatase, R = TiO_2 rutile, T = undetected phase.

Temperature (°C)	Relative weight fraction (%)		Crystallite size (nm)	
	Rutile	Anatase	Rutile	Anatase
200°C	100	—	6 (1)	—
400°C	100	—	11(1)	—
600°C	97.96(5)	2.04(1)	22(1)	98(2)
700°C	99.08(5)	0.92(1)	96(3)	114(6)

Table 5.
Phase weight fraction and TiO_2 crystallite size of the calcined Ti-dissolved powders extracted from XRD data.

XRD peaks imply the nanometric crystal size of the products [28]. MAUD software [26] estimated crystallite size of rutile and anatase in the samples which are presented in **Table 5**.

3.3 Magnesium titanium oxide or magnesium titanate ($MgTiO_3$)

Synthesis of $MgTiO_3$ powder can be performed following the schematic diagram given in **Figure 2**. The detailed explanation of the synthesis has been reported elsewhere [29]. In principle, the magnesium powder was dissolved in hydrochloric acid at room temperature stirring to obtain a light yellow solution, while the titanium powder was prepared similarly but at around 60°C to give a purple solution. Both solutions were then stirred at around 75°C for 4 h to ensure homogenous mixing prior to drying at 105°C to produce granular agglomerates. Phase and microstructure analyses were performed on the manually ground agglomerates after calcination at various temperatures. **Figure 6** shows the XRD patterns of calcined powders for 1:1 Mg:Ti ratio at different temperatures. As can be seen from the figure, only MT (COD ID 01-079-0831, hexagonal), MgO (periclase, 00-045-0946, cubic), and rutile (01-078-2485, tetragonal) were present in all samples. This result shows that our dissolution method can hinder the formation of other magnesium titanate phases such as $MgTi_2O_5$ [30, 31] or Mg_2TiO_4 [32]. However, the

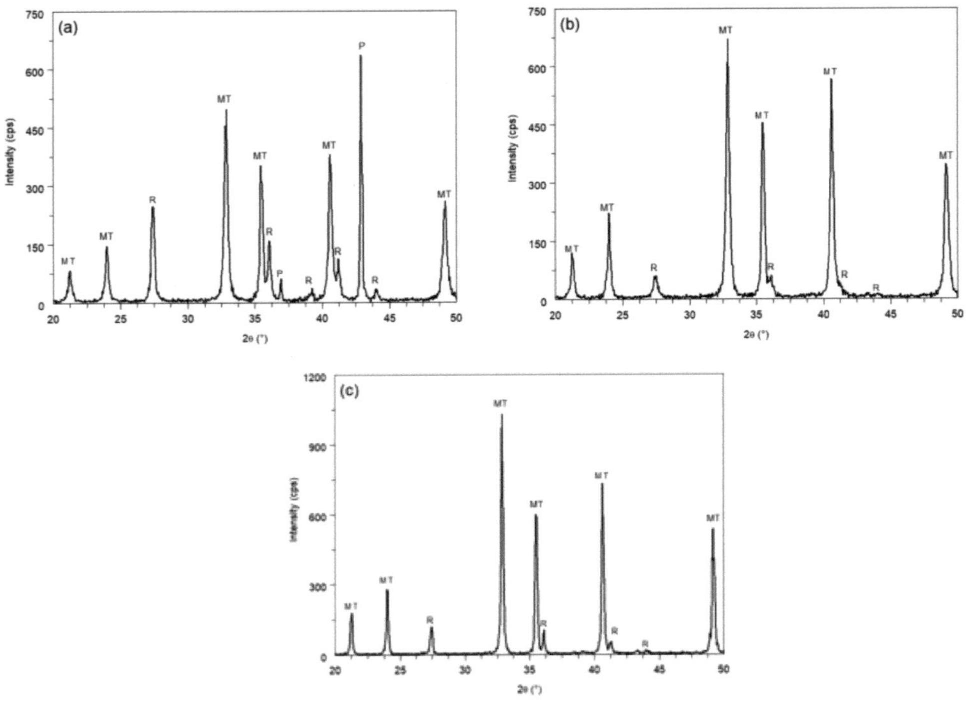

Figure 6.
X-ray diffraction patterns (CuKα radiation) of the calcined (Mg, Ti)-dissolved powders at various calcination temperatures: (a) 600, (b) 700, and (c) 800°C. symbols: MT = $MgTiO_3$, R = rutile, P = periclase (MgO).

presence of periclase and rutile shows that the product is not pure MT. In terms of XRD line peaks, the MT patterns exhibit broadened peaks, particularly at lower temperatures indicating that it is in nanometric crystallite size [17]. The average XRD crystallite size of the MT nanocrystals is between 76 ± 2 nm (600°C) and 150 ± 4 nm (800°C). **Figure** 7 shows a typical TEM micrograph of the MT crystallites. The micrograph shows inhomogeneous crystallite size ranging between 50 and 120 nm—being the dominating size is around 80 nm. Therefore, the TEM result supports that of XRD, where the hexagonal morphology of the crystallites confirms

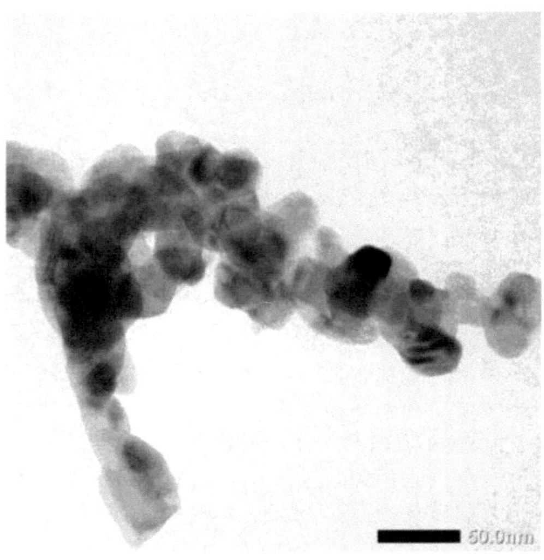

Figure 7.
TEM micrograph of MT powder after calcination at 600°C.

the structure of the phase. In conclusion, the simple dissolution method is very potential to produce bi-cationic nanocrystals.

3.3.1 Effect of excessive Mg

The influence of excessive Mg [33] to produce higher-purity MT was investigated further by XRD for samples which were calcined at 800°C [29]. It has been shown that excessive addition of Mg up to 3% reduces the amount of rutile (R), while above 6% addition causes vanishing rutile but the appearance of periclase (*M*). This result implies that the method can be easily used to control the amount of MT by varying Mg:Ti ratio without the introduction of $MgTi_2O_5$ or Mg_2TiO_4 which was found when other methods were used [30, 32]. The results explained the role of excessive Mg in the formation of high-purity MT. The amount of MT can achieve as high as $99.4 \pm 3.6\%$ in the 6% excessive Mg sample with rutile as the sole residual or $97.5 \pm 3.5\%$ in the 3% excessive Mg sample with periclase as also the only residual. An excessive Mg around 5–6% is therefore envisaged to give MT with the highest purity [29].

3.3.2 Effect of calcination dwelling time

The formation of MT at 600°C with 1 h dwelling time was incomplete, with the weight fraction of rutile and periclase that was about 17.2 and 17.4%, respectively, while increasing calcination temperature improves such formation. The effect of calcination dwelling time on the completeness of the MT formation was then examined [29]. It was reported that the formation of MT is significantly improved by prolonged calcination as proven by the MT molar fraction around 65% (equivalent with around 82% by weight) for 1 h to 82% (90% by weight) for 4 h. It was also found that the crystallite size of MT was invariant to dwelling time at values around 76 nm as indicated by insignificant changes on the XRD line.

Therefore, the dissolution method has been successfully implemented to produce high-purity MT nanocrystals. Sub-nanometric MT crystals were achieved from Mg-Ti hydrochloric acid solutions mixing with 6% excessive Mg followed by calcination at 800°C for 1 h. Meanwhile, prolonged calcination at 600°C significantly improved the MT formation up to 82% (molar) and retained its nanometric crystallite size at around 76 nm.

3.4 Zircon ($ZrSiO_4$)

Synthesis of $ZrSiO_4$ powder was performed by making use of natural zircon sand which was collected from District of Kereng Pangi in Central Kalimantan. To obtain a pure zircon powder, the sand was subject to several processing steps including magnetic separation, reaction with HCl followed with NaOH, and finally washing and drying. The XRD patterns of the samples after each processing are presented in **Figure 8**. Several inferences can be drawn from the figure, for example, the dominating phase in the sand and powder is zircon ($ZrSiO_4$, COD ID 900-2554). Furthermore, the highest intensity increases with the processing step, indicating that the crystallinity improves subsequently. Moreover, the undetected peaks disappear after reacting the sand with HCl. Investigation using XRF showed that the disappearance of the undetected XRD peaks in Sample (B) could be associated with the removal of unwanted substances such as Ti and Fe in the sand (**Figure 9**). However, Sample (B) contains silica quartz (COD ID 500-0035). The XRF data showed Zr:Si weight fraction ratio of approximately 3:2 which indicates that pure zircon powder has not been achieved since the ratio should be at approximately 3:1. Reaction with

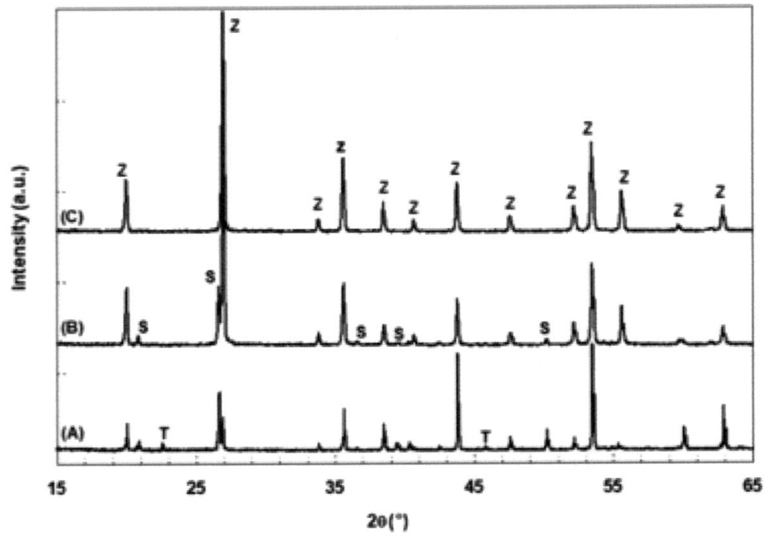

Figure 8.
X-ray diffraction patterns (CuKα radiation) of (A) zircon sand, (B) zircon powder after magnetic separation and reaction with HCl, and (C) powder B after reaction with NaOH and washing. Symbols: Z = zircon, $ZrSiO_4$; S = silica (SiO_2), quartz; T = undetected.

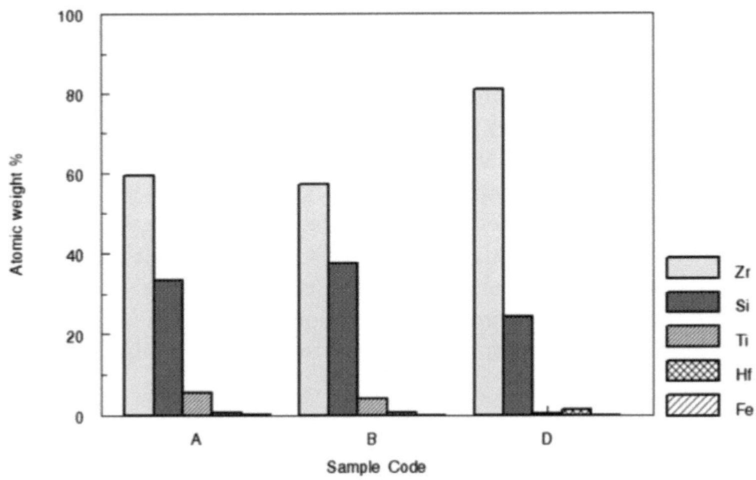

Figure 9.
X-ray fluorescence atomic weight fractions of (a) zircon sand, (B) zircon powder after magnetic separation and reaction with HCl, and (C) powder B after reaction with NaOH and washing.

NaOH was subsequently done to form liquid Na_2SiO_3 precursor which can be easily removed by sieving. The sieved slurry was then washed and dried to obtain pure zircon powder (Sample (C)). The SEM image and its energy-dispersive elemental mapping of Sample C are shown in **Figure 10**. The figures indicate the high crystallinity and high purity of the powder. These facts show that the dissolve method which comprises magnetic separation, HCl reaction, washing-filtering, NaOH reaction, another washing-filtering, and drying of natural zircon sand may produce high-purity, single-phase zircon powder.

An effort to process the zircon powder to achieve zircon nano-powder has been accomplished. The mechanical zirconia ball-milling was chosen. Well-sintered partially stabilized zirconia ceramics exhibit much higher Vickers' hardness than zircon

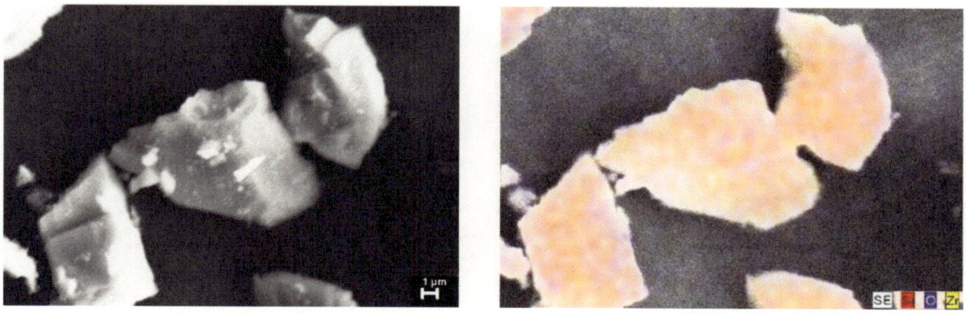

Figure 10.
SEM image (SEI mode, left) and its energy-dispersive elemental mapping (right) of the purified zircon powder (sample C).

ceramics, that is, 15 GPa [34] versus 3 GPa [35]. Therefore, zirconia balls would presumably be able to mill zircon powder. The XRD patterns of the milled zircon powders at various milling time are presented in **Figure 11**. Milling has no effect on the detected phase, that is, it remains zircon, and resulted in lower but broader peak intensity. The XRD peak broadening was caused by crystallite size and nonuniform effects after milling. These XRD characters were investigated using MAUD software [26] and came across at a conclusion that the synthesized zircon powder exhibited smaller crystallite size after milling, that is, from 173 nm before milling to 162 and finally to 45 nm after milling for 5 and 10 h, respectively. TEM images of the unmilled and 10 h milled zircon powders are shown in **Figure 12**. These images confirm the success of the synthesis of the nano-sized zircon powder.

The software also allowed the extraction of a nonuniform strain of the milled powder, and the results showed that longer milling induced more strains as also observed by others [36, 37] for different metal oxides. The strain values for the zircon powders are 2, 3, and 6×10^{-4} for the unmilled and 5 and 10 h for milled samples, respectively. These values dropped significantly to as low as 2×10^{-4} after annealing at 200°C for 2 h.

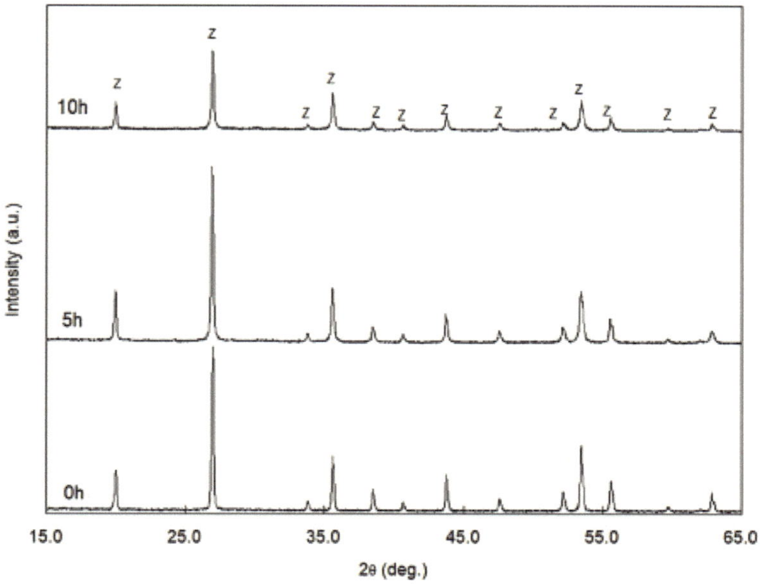

Figure 11.
X-ray diffraction patterns (CuKα radiation) of milled zircon powder at various milling times. Symbol: Z = zircon, $ZrSiO_4$.

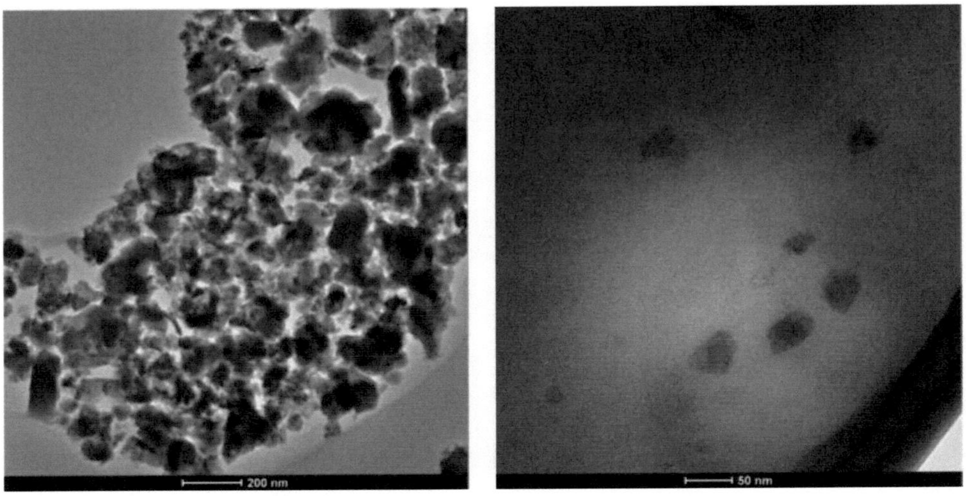

Figure 12.
TEM micrographs of unmilled (left) and 10 h milled zircon (right) powders.

4. Conclusion

It has been demonstrated that a simple dissolve method using HCl or NaOH which was reacted with a mono- or bi-cationic metals or with a natural mineral can be accomplished to produce high-purity, nanocrystalline metal oxide ceramics. Pure MgO, TiO_2, $MgTiO_3$, and $ZrSiO_4$ nano-powders were produced by the method. The main step for the metal case was on the formation of the precursor and appropriate calcination temperature. For the mineral case, purification of the phase under consideration by reacting the contaminating phase with acid or base, followed by filtering, washing, and calcination, was the key of the process.

Acknowledgements

The authors express gratitude to Kemenristekdikti of the Republic of Indonesia and LPPM ITS who support the research funding through PBK-2017 No. 528/PKS/ITS/2017.

Author details

Suminar Pratapa[1*], Ella A.D. Kiswanti[1], Dien R. Diana[1], Yufi Hariyani[1], Lisma D.K. Sari[2], Musyarofah Musyarofah[1], Triwikantoro Triwikantoro[1] and Malik A. Baqiya[1]

1 Department of Physics, Faculty of Science, Institute of Technology Sepuluh Nopember (ITS), Surabaya, Indonesia

2 Study Program of Mathematics, STKIP PGRI, Situbondo, Indonesia

*Address all correspondence to: suminar_pratapa@physics.its.ac.id

References

[1] Li T, Liao JP, Wang YF. Solvothermal synthesis and magnetic properties of β-Co_2P nanorods. Materials Science-Poland. 2015;**33**:312-316. DOI: 10.1515/msp-2015-0049

[2] Sin J-C, Lam S-M, Lee K-T, Mohamed AR. Surfactant-free solvothermal synthesis of ZnO nanorods for effective sunlight degradation of 2,4-dichlorophenol. Materials Letters. 2015; **140**:51-54. DOI: 10.1016/j.matlet.2014.10.067

[3] Chun J, Jo C, Lim E, Roh KC, Lee J. Solvothermal synthesis of sodium cobalt fluoride ($NaCoF_3$) nanoparticle clusters. Materials Letters. 2017;**207**:89-92. DOI: 10.1016/j.matlet.2017. 07.059

[4] Kurajica S, Minga I, Grčić I, Mandić V, Plodinec M. The utilization of modified alkoxide as a precursor for solvothermal synthesis of nanocrystalline titania. Materials Chemistry and Physics. 2017;**196**:194-204. DOI: 10.1016/j.matchemphys.2017.04.064

[5] Tamilselvi P, Yelilarasi A, Hema M, Anbarasan R. Synthesis of hierarchical structured MgO by sol-gel method. Nano Bulletin. 2013;**2**:130106

[6] Soleimanian V, Aghdaee SR. X-ray diffraction analysis of the effect of annealing temperature on the microstructure of magnesium oxide nanopowder. Journal of Physics and Chemistry of Solids. 2015;**81**:1-9. DOI: 10.1016/j.jpcs.2014.12.020

[7] Jeevanandam J, Chan YS, Danquah MK. Calcination-dependent morphology transformation of sol-gel-synthesized MgO nanoparticles. ChemistrySelect. 2017;**2**:10393-10404. DOI: 10.1002/slct.201701911

[8] Dubey RS. Temperature-dependent phase transformation of TiO_2 nanoparticles synthesized by sol-gel method. Materials Letters. 2018;**215**:312-317. DOI: 10.1016/j.matlet.2017.12.120

[9] Shi Y, Huang X, Yan D. Synthesis and characterization of ultrafine zircon powder. Ceramics International. 1998;**24**:393-400. DOI: 10.1016/S0272-8842(97)00027-8

[10] Ran FY, Cao WB, Li YH, Zhang XN. Preparation of nanosize anatase TiO_2 powders by hydrothermal synthesis. Key Engineering Materials. 2007;**336–338**:2017-2020. DOI: 10.4028/www.scientific.net/KEM.336-338.2017

[11] Rajendran V, Deepa B, Mekala R. Studies on structural, morphological, optical and antibacterial activity of pure and Cu-doped MgO nanoparticles synthesized by co-precipitation method. Materials Today: Proceedings. 2018;**5**:8796-8803. DOI: 10.1016/j.matpr.2017.12.308

[12] Pei L-Z, Yin W-Y, Wang J-F, Chen J, Fan C-G, Zhang Q-F. Low temperature synthesis of magnesium oxide and spinel powders by a sol-gel process. Materials Research. 2010;**13**:339-343. DOI: 10.1590/S1516-14392010000300010

[13] Kamarulzaman N, Chayed NF, Badar N. MgO nanoparticles via a simple solid-state reaction. In: International Symposium on Frontier of Applied Physics (ISFAP) 2015; 5-7 October 2015; Bandung: AIP Publishing. 2016;**1711**: p. 040004. DOI: 10.1063/1.4941626

[14] Cai W, Yang H, Guo X. A facile synthesis of nanocrystalline spherical TiO_2 particles and its photoluminescent properties. Procedia Engineering. 2014;**94**:71-75. DOI: 10.1016/j.proeng.2013.11.042

[15] Nachit W, Touhtouh S, Ramzi Z, Zbair M, Eddiai A, Rguiti M, et al. Synthesis of nanosized TiO_2 powder by sol gel method at low temperature. Molecular Crystals and Liquid Crystals. 2016;**627**:170-175. DOI: 10.1080/15421406.2015.1137135

[16] Parthasarathy G, Manorama SV. A novel method for synthesizing nano-crystalline $MgTiO_3$ geikielite. Bulletin of Materials Science. 2007;**30**:19-21. DOI: 10.1007/s12034-007-0004-y

[17] Kanna RR, Dhineshbabu NR, Paramasivam P, Rajendran V, Yuvakkumar R. Synthesis of geikielite ($MgTiO_3$) nanoparticles via sol–gel method and studies on their structural and optical properties. Journal of Nanoscience and Nanotechnology. 2016;**16**:7635-7641. DOI: 10.1166/jnn.2016. 11114

[18] Tamin SH, Adnan SBRS, Jaafar MH, Mohamed NS. Effects of sintering temperature on the structure and electrochemical performance of Mg_2SiO_4 cathode materials. Ionics. 2017; **24**:2665-2671. DOI: 10.1007/s11581-017-2391-4

[19] Roald B, Beck W. The dissolution of magnesium in hydrochloric acid. Journal of the Electrochemical Society. 1951;**98**: 277. DOI: 10.1149/1.2778207

[20] Wellens S, Vander Hoogerstraete T, Möller C, Thijs B, Luyten J, Binnemans K. Dissolution of metal oxides in an acid-saturated ionic liquid solution and investigation of the back-extraction behaviour to the aqueous phase. Hydrometallurgy. 2014;**144–145**:27-33. DOI: 10.1016/j.hydromet.2014.01.015

[21] Casey WH, Ludwig C. The mechanism of dissolution of oxide minerals. Nature. 1996;**381**:506-509. DOI: 10.1038/381506a0

[22] Pinto AH, Souza FL, Chiquito AJ, Longo E, Leite ER, Camargo ER. Characterization of dense lead lanthanum titanate ceramics prepared from powders synthesized by the oxidant peroxo method. Materials Chemistry and Physics. 2010;**124**: 1051-1056. DOI: 10.1016/j.matchemphys.2010.08.030

[23] Shabashov VA, Sagaradze VV, Litvinov AV, Mukoseev AG, Vildanova NF. Mechanical synthesis in the iron oxide—Metal system. Materials Science and Engineering A. 2005;**392**:62-72. DOI: 10.1016/j.msea.2004.11.006

[24] Reutov DS, Khalezov BD, Ovchinnikova LA, Gavrilov AS. Investigation of zinc ferrite dissolution kinetics by roll-disc method. Tsvetnye Metally. 2017;**11**:12-15. DOI: 10.17580/tsm.2017.11.02

[25] Hunter BA. Rietica. International Union of Crystallography Commission on Powder Diffraction Newsletter. 1998; **20**:21

[26] Lutteroti L. MAUD: Material Analysis Using Diffraction. Availble from: http://maud.radiographema.eu/ [Accessed: Aug 25, 2018]

[27] Samuel V, Pasricha R, Ravi V. Synthesis of nanocrystalline rutile. Ceramics International. 2005;**31**:555-557. DOI: 10.1016/j.ceramint.2004.07.003

[28] Pratapa S, Susanti L, Insany YAS, Alfiati Z, Hartono B, Mashuri A, et al. XRD line-broadening characteristics of M-oxides (M = Mg, Mg-Al, Y, Fe) nanoparticles produced by coprecipitation method. In: Proceedings of the Third Nanoscience and Nanotechnology Symposium (NNSB2010); 16 June 2010; Bandung: AIP Publishing. 2010;**1284**:125-128

[29] Pratapa S, Baqiya MA, Istianah I, Lestari R, Angela R. A simple dissolved metals mixing method to produce high-purity $MgTiO_3$ nanocrystals. In: Proceedings of the Fifth Nanoscience

and Nanotechnology Symposium (NNS2013); 23-25 October 2013; Surabaya: AIP Publishing. 2014;**1586**: 39-42. DOI: 10.1063/1.4866726

[30] Bernard J, Belnou F, Houivet D, Haussonne J-M. Synthesis of pure $MgTiO_3$ by optimizing mixing/grinding condition of $MgO + TiO_2$ powders. Journal of Materials Processing Technology. 2008;**199**:150-155. DOI: 10.1016/j.jmatprotec.2007.07.044

[31] Tang B, Zhang S, Zhou X, Deng C, Yu S. Preparation of pure $MgTiO_3$ powders and the effect of the $ZnNb_2O_6$-dope onto the property of $MgTiO_3$-based ceramics. Journal of Alloys and Compounds. 2010;**492**:461-465. DOI: 10.1016/j.jallcom.2009.11.140

[32] Miao Y-M, Zhang Q-L, Yang H, Wang H-P. Low-temperature synthesis of nano-crystalline magnesium titanate materials by the sol-gel method. Materials Science and Engineering B. 2006;**128**:103-106. DOI: 10.1016/j.mseb.2005.11.019

[33] Sreedhar K, Pavaskar NR. Synthesis of $MgTiO_3$ and $Mg_4Nb_2O_9$ using stoichiometrically excess MgO. Materials Letters. 2002;**53**:452-455. DOI: 10.1016/S0167-577X(01)00525-0

[34] Hirano M, Inada H. Fracture toughness, strength and Vickers hardness of yttria-ceria-doped tetragonal zirconia/alumina composites fabricated by hot isostatic pressing. Journal of Materials Science. 1992;**27**: 3511-3518. DOI: 10.1007/BF01151827

[35] Rendtorff NM, Grasso S, Hu C, Suarez G, Aglietti EF, Sakka Y. Dense zircon ($ZrSiO_4$) ceramics by high energy ball milling and spark plasma sintering. Ceramics International. 2012; **38**:1793-1799. DOI: 10.1016/j.ceramint.2011.10.001

[36] Indris S, Bork D, Heitjans P. Nanocrystalline oxide ceramics prepared by high-energy ball milling. Journal of Materials Synthesis and Processing. 2000;**8**:245-250. DOI: 10.1023/A:1011324429011

[37] Ekström T, Chatfield C, Wruss W, Maly-Schreiber M. The use of X-ray diffraction peak-broadening analysis to characterize ground Al_2O_3 powders. Journal of Materials Science. 1985;**20**: 1266-1274. DOI: 10.1007/BF01026322

Chapter 4

Plasma Resistance Evaluation and Characteristics of Yttria Ceramics Sintered by Using Calcination Yttria

Jin Sam Choi

Abstract

The evaluation of plasma resistance and the characteristics of yttria ceramics fabricated by calcination yttria as a starting material without dopants under an oxidation atmosphere were investigated. Regardless of the starting materials, as-received and calcined yttria powder, XRD patterns showed that all samples have Y_2O_3 phase. The three cycling processes inhibited a large grain, which occurs frequently during the yttria sintering, and a high-density ceramic with a homogeneous grain size was obtained. The smaller the grain size, the larger were the Young's modulus and K_{IC}. Compared to Al_2O_3 and ZrO_2 ceramics, yttria ceramics showed 3 times larger plasma resistance and 1.4–2.2 times lower weight loss during the plasma etching test, respectively. The characteristics of pure hot-pressed yttria ceramics tempered in an oxidation atmosphere are also investigated. Even though the color variation of the hot-pressed Y_2O_3 ceramics was due to the sintering temperatures, the oxidation process turned the color of the Y_2O_3 ceramics into white. In addition, oxygen defects also affected the weight change and microstructure of the Y_2O_3 ceramics. The Y_2O_3 ceramic sintered at 1600°C had obtained a full density. As the sintering temperature increased, small homogeneous grains grew to large grains which affected the Vickers hardness.

Keywords: as-received yttria, calcined yttria, three-repeated cycle sintering, homogeneous grain, plasma resistance, hot-pressed yttria ceramics, oxidation, homogeneous grain, full density

1. Approach to sintering a yttria monolith to full density by surface and interfacial engineering

1.1 Introduction

Yttria, Y_2O_3, has great potential as a host material for solid-state lasers such as yttrium aluminum garnet-doped neodymium, Nd:YAG, and yttrium oxide doped with ytterbium, Yb:Y_2O_3 [1, 2]. Other commonly known oxides, such as Al_2O_3, MgO, and ZrO_2 ceramics, have attracted more interest than yttria [3, 4]. Therefore, the application of yttria ceramics in modern industry has been limited. Yttria ceramics are strongly dependent on not only their intrinsic properties but also their crystal structure. Yttria, which has a C-type rare earth sesquioxide structure, needs to be derived from the cubic fluorite-type structure by removing one quarter of the oxygen atoms [5]. Yttria has 32 yttrium and 48 oxygen ion sites per unit cell. This structure has large interstitial sites with the same size as an oxygen ion in the anion sublattice. Yttria has a cubic or alpha-type crystal structure until the temperature reaches 2325°C. The phase transition to a tetragonal crystal structure also occurs at 2325°C. A stable phase, known as hexagonal or beta-type, is maintained until it reaches the melting temperature, 2340°C [6].

Recently, yttria, which is used widely in partially stabilized zirconia and sintering aids, was the center of attention in the semiconductor industry because yttria is superior to quartz, Al_2O_3, ZrO_2, BN, and SiC in terms of its radical or cationic resistance activated by plasma [7] and sintering-limiting property [6].

Nevertheless, when the synthesis of nanoparticles using coprecipitation was carried out and the influence of precursors or additives was examined, it was reported that yttria, sintered under vacuum or hydrogen conditions, was close to the theoretical density [8–10]. The additives affected the grain boundary migration. This compensated for the charge, cation diffusion, solute transport, and additive-defect interaction [11, 12]. According to Huang et al. [13], the theoretical density of yttria ceramic was obtained by the two-step sintering and vacuum sintering of a lanthanum-doped yttria ceramic combination. Previous studies [14–16] attempted to improve the yttria transmittance and electrical conduction of the body sintered using special techniques, such as spark plasma sintering, microwave-flash combustion synthesis, hot isostatic pressing, and the addition of tri- or tetravalent additives. Therefore, it is necessary to examine the intrinsic sintering characteristics of yttria itself. More recently, Choi et al. [17] reported not only the behavior, color, and density of yttria ceramics but also the weight change due to oxygen vacancies and oxygen diffusion in a sintered body as a function of the sintering temperature. They suggested that the changes in color, density, weight, and microstructure of grains according to the sintering temperature were related to the volatilization of yttrium ions at oxygen vacancies in the lattice site at high temperatures [8, 10, 11]. On the other hand, the sintering property of the starting material, calcined yttria, from which the hydration reaction had been eliminated, as expressed in Eq. (1), but oxygen diffusion in oxygen vacancies and yttria powder, which removes gas spouting of the precursor, occurred continuously up to 1250°C, is unknown [13, 15]:

$$RE\,(H_2O)_n^{3+} + H_2O\left[RE(OH)\,(H_2O)_n^{-1}\right]^{2+} + H_3O^+ \qquad (1)$$

where RE and $_n$ represent the coordination number of rare earth and cation, respectively.

To determine the effects of oxygen vacancies and hydration on the sintering properties of yttria, this study studied the sintering characteristics of yttria calcined without a gas or hydration reaction containing a hydration reagent of precursors. In addition, when oxygen diffusion occurred in the oxygen vacancies, the yttria powder was heated repeatedly to adopt the result of density and increasing weight. The plasma resistivity of the yttria ceramic was compared with that of the control group.

1.2 Y_2O_3 as a host material and fabrication methods for the potential applications

1.2.1 Y_2O_3 structure and dopant

Y_2O_3 has a C-type cubic structure of rare earth oxides. The C-type structure is due to the bixbyite (space group Ia3 (Th7) with X = 16) and is similar to the fluorite

(CaF_2)-type structure, in which there is a quarter of empty anion sites and they are arranged regularly (**Figure 1**).

Three quarters of the cations are in cubes, in which the missing oxygen ions are along the face diagonal. One quarter of the cations are in cubes, in which the missing oxygen ions are along the body diagonal. The unit cell contains 48 oxygen and 32 yttrium ions. Yttrium ions are located at the center of an approximate cube, and oxygen ions are located at six of the eight vertices [1, 18]. In three quarters of these sites, the missing oxygen ions are located on a face diagonal of the mini-cube, and the rest are located at the ends of a body diagonal. The full unit cell contained 4 × 4 × 4 of these mini-cubes. Here, only a 2 × 2 × 1 part is shown for clarity of presentation. As an oxide, yttria showed broad transmittivity from 280 nm to 8 μm. Y_2O_3 melts at 2430°C, but the C-type structure was only stable up to 2325°C [6]. The H-type structure has a hexagonal form at high temperatures. The high melting temperature and this polymorphic transition make the growth of yttria single crystals difficult. Like fluorite-structured oxides, oxygen vacancies and interstitials are the major defects in pure yttrium oxide [19].

In fluorite-structured oxides, Y_2O_3 can dissolve a large amount of aliovalent cations. This is accompanied by the formation of charge-compensation oxygen vacancies when accepted dopants are present and donor dopants are present in the oxygen interstitials. These anion defects facilitate oxygen diffusion. Indeed, oxygen anion diffusion was much faster than yttrium cation diffusion in pure Y_2O_3. Therefore, cation diffusion is the rate control step for grain boundary migration at all compositions. Because cation doping can be conducted readily within the solubility limit, Y_2O_3 like CeO is a good candidate for examining the cation dopant effects on grain boundary mobility [18]. The cubic symmetry of this oxide further ensured a lack of strong anisotropy in grain boundary mobility, which could otherwise complicate grain growth behavior.

Dopants of both donor and acceptor types as additives for sintering yttria at lower temperatures include Mg^{2+}, Sr^{2+}, Sc^{3+},Yb^{3+}, Gd^{2+}, La^{3+}, Ti^{4+}, Zr^{4+}, Ce^{4+}, and Nb^{5+} [19, 20]. The role of the dopant in yttria is similar to fluorite-structured oxide estimated from several important features. The grain boundary mobility by cation diffusion involves an interstitial mechanism. Solute drag could suppress grain boundary mobility at high dopant concentrations. Grain boundary mobility is influenced by a dopant-defect interaction, which is charge and size dependent. Moreover, undersized dopants have a tendency to enhance grain boundary mobility markedly due to distortion of the surrounding lattice that apparently facilitates

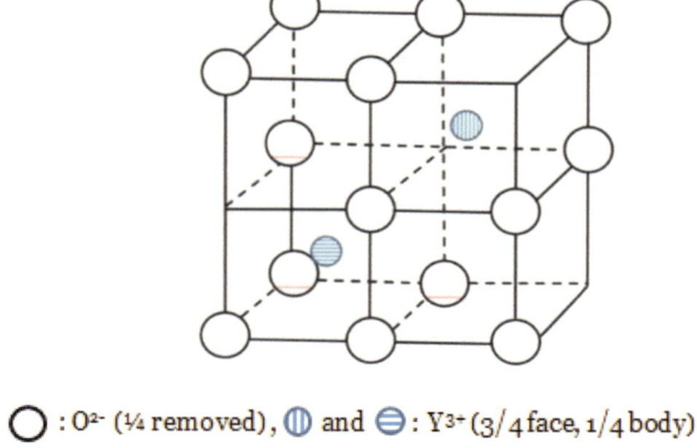

Figure 1.
Crystal structure of Y_2O_3.

defect migration [12, 18]. According to Kingery et al. [21], the diffusion of oxygen anions in yttria is much faster than that of yttrium cations. Hence, yttrium cation interstitial diffusion is the rate control step for grain boundary migration.

1.2.2 Yttria ceramic fabrication

Yttria has potential applications, such as missile domes and bulb envelopes because of its optical transparency and a host material for solid-state lasers and high-resistance materials for etching plasma. Several studies have focused on obtaining full-density or high-transparency polycrystalline yttria ceramics. In the former, several methods have attempted to control the particle size and its distribution, coprecipitation, dopants or additives, and homogeneity, through hydrothermal methods. In the latter, a novel sintering method was studied. A combination of vacuum sintering and hot isostatic pressing methods was based on the low-temperature sintering process to suppress grain growth without sacrificing densification. The pores can be removed completely after hot isostatic pressing. These results are based on the addition of additives in yttria. Ikegami et al. [22] reported the fabrication of transparent ceramics by the two-temperature sintering of yttrium hydroxide with near theoretical total transmittance. Haung et al. [13] synthesized lanthanum-doped yttria ceramic by a combination of two-step sintering and vacuum sintering. A two-step strategy was proposed to produce materials with near theoretical total transmittance, and these studies focused on transparent ceramics used as infrared windows.

1.3 Sample fabrication

The sample preparation was as follows. Yttria powder (99.9%, UUHP, Shin-Etsu Chemical Co. Ltd., Japan) was used as the starting material of as-received yttria. ASY and CY powder was prepared by calcination at 1250°C for 48 h. The powder of fused yttria, FY (Mirae Materials Co. Korea), by induction melting was ball milled with 300 μm zirconia ball media for 24 h. After milling, the powder mixtures were dried at 500°C for 24 h and sieved with 325 mesh to achieve uniform particle sizes. Here, the graphite mold having a diameter of 60 mm was filled with 100 g of the starting material. After forming one axis at 200 kg, cold isostatic pressing (CIP) was performed at a 300 bar pressure to increase the green density. The heating and cooling rates were 5°C/min. The sintered specimens listed in **Table 1** were obtained by the thermal history shown in **Figure 2**.

In the crystal phase, a coarse grain microstructure, and a microstructure of specimen, the specimens were observed using X-ray diffraction (XRD, D/MAX-2550V, Rigaku, Japan), polarizing microscope (Eclipse LV100, Nikon, Japan), and scanning electron microscope (SEM, JSM-6700F, Jeol, Japan), respectively, while bulk density of the sintered specimens was measured according to the ASTM D6683-14 method. The modulus of elasticity of the sintered specimens was observed using a pulse echo device (5800 Pulser/Receiver, Panametrics, Japan) with an oscilloscope (Tektronix, TDS3012, Japan) of X- and Y-modulation, longitudinal, and transverse waves. K_{IC} (HV-100, Mitutoyo, Japan) was measured by applying a load of 500 g on average of 10 times or more after the specimen surface was polished with 0.2 μm diamond paste. The etching depth and the weight reduction rate were obtained by surface roughness (Surfcorder ET3000, Kosaka Lab., Japan) after plasma etching for 60 min by using an inductively coupled plasma system (Versiline, UNAXIS Co., USA). Sintering properties and plasma evaluation were compared with yttria ceramics sintered, Al_2O_3 (99.6%, Semiconductor Wafer, Inc., Taiwan), ZrO_2 (Semiconductor Wafer, Inc., Taiwan), Si wafer (Sumco Co., Japan), and quartz (General Electric (GE), USA), respectively.

Name	Starting materials
ASY	As-received yttria
CY	Calcined yttria, at 1250°C for 24 h in air
FY	Fused yttria by induction melting
HY [17]	Hot-pressed, at 1600°C for 8 h under Ar gas

Table 1.
Yttria ceramics named and prepared.

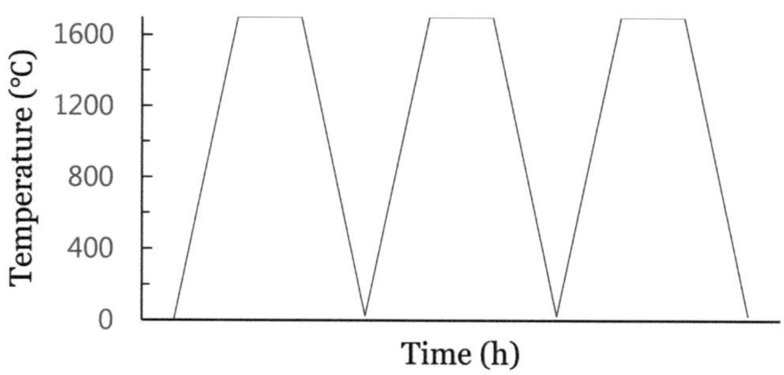

Figure 2.
Sintering schedule of the three-repeated cycle in air.

1.4 Characteristic evaluation of yttria ceramics sintered by three-repeated cycle

Table 1 lists the names of yttria ceramic specimens. **Figure 3** shows the XRD crystal phase of the yttria ceramic sintered under the conditions in **Figure 2**. X-ray diffraction analysis (XRD, D/MAX-2550V, Rigaku, Japan) was conducted to determine the crystal structure of the prepared yttria ceramics sintered using CuKα1 radiation between 5 and 70° (2θ) at room temperature. The crystal phase of CY, the calcined starting material, was identified as Y_2O_3, but ASY contained both Y_2O_3 and $YO_{1.335}$.

The difference in the crystallographic phase depending on the starting material is believed to be the effect of the oxygen vacancies, which was about one-fourth of the 48 standard positions [5, 11, 15].

The starting material calcined at 1250°C for 48 h was exposed to hydration and precursor gases, whereas ASY exhibited continuous volatilization of precursors up to 1250°C [10, 16]. Rhodes et al. [23] examined the continuous volatilization of methane gas. They observed not only the release of carbon or water impurities but also carbonization and hydration reactions until the temperature reached 1200°C. At higher temperatures, the calcined starting material was found to be free from the influence of the adsorption gas and hydration reaction of the precursor [5, 9, 10, 16]. A mixture phase of Y_2O_3 and $YO_{1.335}$ in ASY was obtained after the sintering, and the relative oxygen filling was less than CY [13]. Based on this, similar study results were reported; the self-diffusion coefficient and activation energy at 1050–1250°C were 6×10^{-6} cm^2/s and 82 kJ/mole, respectively [24]. Swamy et al. [25] suggested the possibility of a fluorite crystal phase formation due to the irregular oxygen vacancies in the unit lattice according to high-temperature XRD equipment.

In addition, there was a difference between the results in that the additive affected only the change in transmittance, $YO_{1.401}$ or $YO_{1.335}$ phase formation, and the crystalline phase of the monoclinic were mixed depending on the heat treatment conditions. On the other hand, these results are common in that it originated from oxygen vacancies in

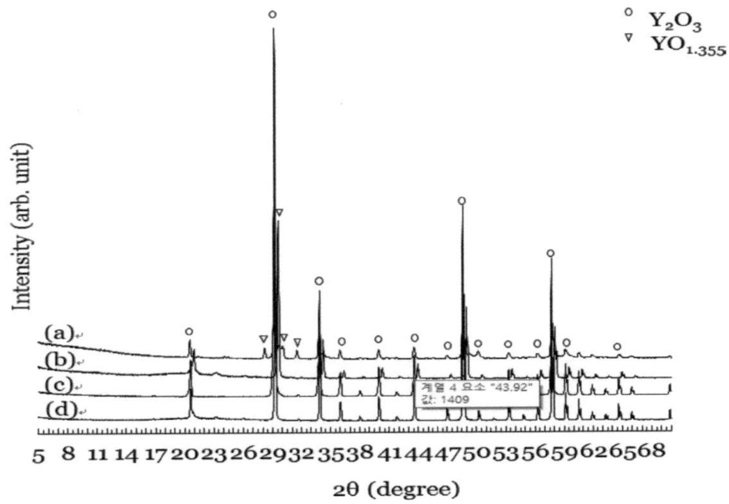

Figure 3.
XRD patterns of yttria ceramics sintered. (a) ASY, as-received yttria; (b) CY, calcined yttria; (c) FY, fused yttria; and (d) HY [17], hot-pressed yttria.

the yttria regular site [13, 18, 24]. The density of the CY specimen was 4.9 g/cm^3, which is higher than that of ASY, 4.8 g/cm^3, as shown in **Figure 4**. On the other hand, CY, calcined yttria, and ASY, repeated-cycle sintering yttria, corresponding to the density of commercial yttria (Y0100A, Kyocera, Japan) and 4.9 g/cm^3 were effective sintering methods, despite being lower than the theoretical density (5.01 g/cm^3) of the HY specimen.

Figure 5 shows a coarse grain microstructure of the CY specimen, in which the surface of the specimen was polished with a 0.2 μm diamond paste, regardless of the starting material condition. The microstructure after repeated-cycle sintering revealed a large number of open pores, in which the hydration reactant and precursor gas of formula (1) were observed, but no coarse particles were observed in the ASY specimen in **Figure 6(a)** [9, 24].

In addition, the CY specimen showed a relatively dense microstructure, as shown in **Figure 6(b)**. The coarse particles in the initial stage changed to uniform particles after repeat-cycle sintering of yttria in air (see **Figures 5** and **6(a)**). Because Y_2O_3 is an oxygen-deficient complex, it is easy to deviate from the stoichiometric Y:O ratio of 40:60 (at.%) due to oxygen vacancies [13, 25]. The majority

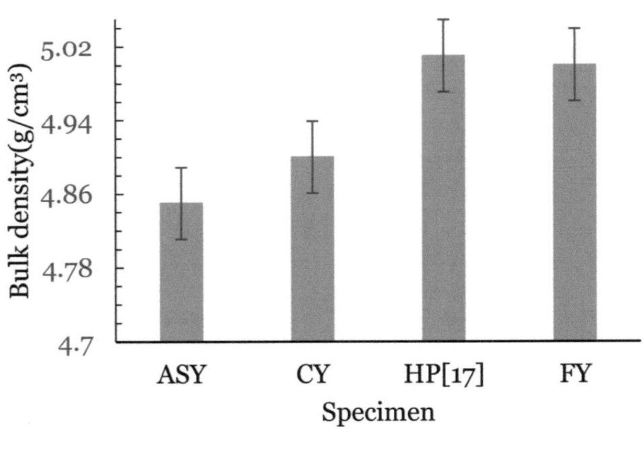

Figure 4.
Bulk density of yttria ceramics sintered.

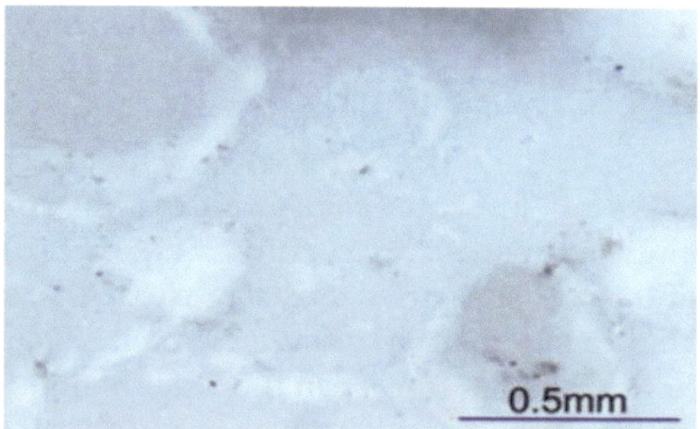

Figure 5.
Optical photograph of the polished surface of CY, yttria ceramics, the first step sintered at 1700°C for 6 h in air. Different granules in the matrix represent abnormal grain growth. The black dots show the remaining polishing powders.

phase of $YO_{1.335}$ with oxygen deficiencies appeared in only the ASY specimen observed in XRD patterns in **Figure 3**.

A thermodynamically stable liquid phase formed locally at 1522°C when the Y:O ratio was 60:40 (at.%), which resulted in pyrolysis of the liquid phase in the repeated heat treatment [25]. The reasons why the rough particles of **Figure 5**, which appeared at the initial stages of less-pressure sintering of yttria, changed to homogeneous particles like the microstructure in **Figure 6(a)** after repeated sintering are as follows. As Y_2O_3 is an oxygen-deficient body, it is easy to deviate from the stoichiometric Y:O ratio range of 40:60 (at.%) due to oxygen vacancies [13]. In **Figure 3(a)**, which showed the XRD crystal phase, only the $YO_{1.335}$ crystal phase with oxygen deficiencies appeared in the ASY specimen. The liquid phase was thermodynamically stable, formed locally at 1522°C and decomposed thermally in the repeated heat treatment, resulting in a crystal when the Y:O ratio was in 60:40 (at.%). The coarse particles in **Figure 5** were prepared by the liquid phase formation when the Y:O ratio was 60:40 (at.%) at 1522°C.

This liquid phase changed stoichiometrically at a Y:O ratio of 40:60 (at.%) due to oxygen diffusion in the lattice during repeated sintering. Therefore, the coarse particles in **Figure 4** disappear because it could change into a crystalline phase rather than exist in the liquid phase [25, 26]. **Figure 7** shows the Y-O phase diagram at 1522°C when the Y:O ratio was 60:40 (at.%); coarse particles support the validity of liquid phase formation hypothesis. This liquid phase changed stoichiometrically at a Y:O ratio (at.%) of 40:60 due to oxygen diffusion in the lattice during the repeated

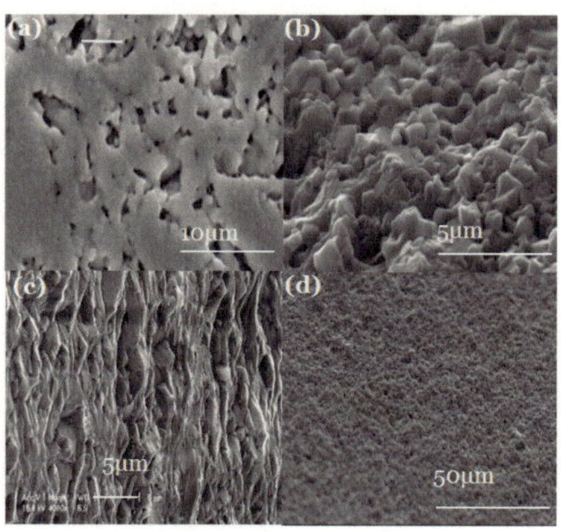

Figure 6.
SEM images of (a) ASY, as-purchased Y_2O_3 powder; (b) CY, calcined Y_2O_3 powder at 1250°C for 48 h in air; (c) fused yttria by induction melting; and (d) HY, hot-pressed yttria [17].

sintering. Therefore, it is believed that the coarse particles shown in **Figure 4** disappear because the crystalline phase was more stable than the liquid phase, but this hypothesis will be proven through further studies. **Figure 8(a)** and **(b)** showed the Vickers indentation results of the ASY and CY specimens. Stress propagation along the cleavage plane and the low-level mist shape were distinct from the HY specimen, which was similar to the glass-like microstructure in **Figure 8(c)**.

The Young's modulus increased with decreasing particle size, as listed in **Table 2**, where average values of the indentation of the specimens, K_{IC}, were measured to apply a 500 g load more than ten times. On the other hand, the HY specimen K_{IC} appeared to be relatively high regardless of the particle size because the dense microstructure, crystal phase, and the amorphous phase were mixed as shown in **Figures 6(c)** and **8(c)** [26, 27]. **Figure 9** presents the plasma resistivity profiles of the sintered specimens and the control group. From Eqs. (2) and (3), the etching depth and weight reduction rate were obtained by surface roughness after plasma etching for 60 min using an inductively coupled plasma system by masking the abrasive specimen.

The sintering properties and plasma evaluation of yttria ceramic were compared with Al_2O_3 (Semiconductor Wafer, Inc., Taiwan), ZrO_2 (Semiconductor Wafer, Inc., Taiwan), Si wafer (Sumco Co., Japan), and quartz (General Electric) [11]. The etch rate and weight loss rate, which are related directly to the plasma resistance, were calculated as follows. The evaluation of plasma etching was performed using simple equations:

$$\text{Etch rate } (\mu\text{m min}) = \text{etch depth } (\mu\text{m})/\text{exposure time (min)} \qquad (2)$$

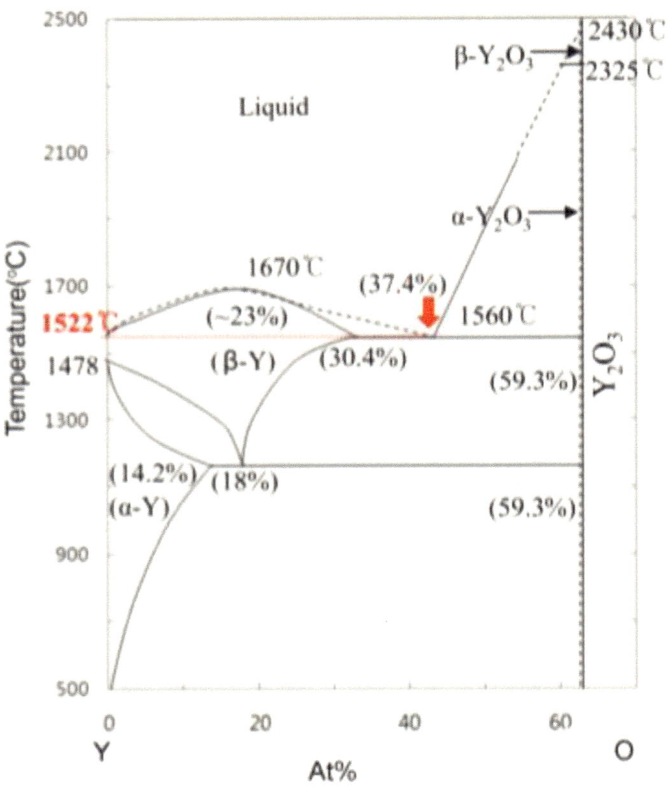

Figure 7.
Phase diagram of yttrium-oxygen [6].

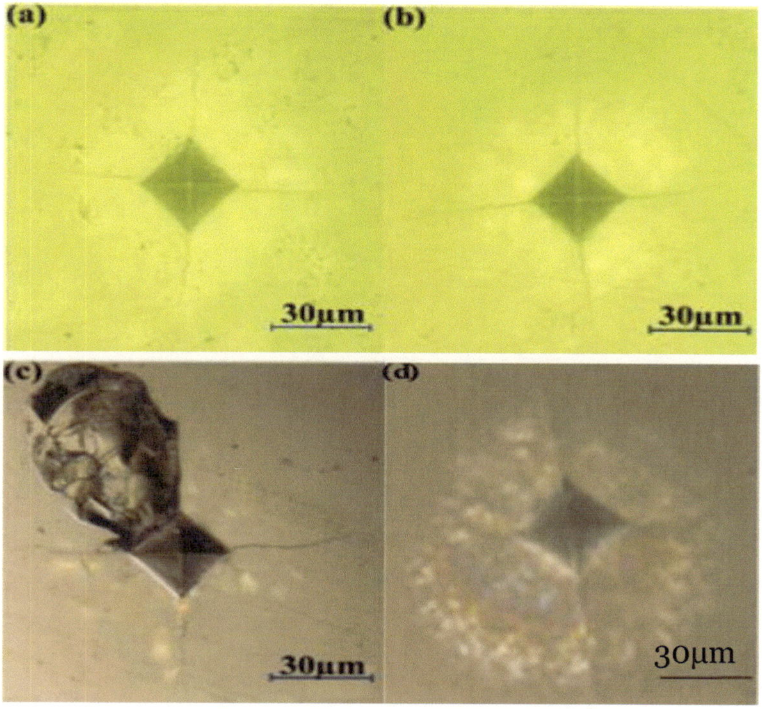

Figure 8.
Vickers indents in sintered polycrystalline yttria ceramics of varying sinter method at 500 g load. Note the general tendency for both more cracking and greater complexity of cracking, particularly near the indent: (a) ASY, as-received yttria; (b) CY, calcined yttria; (c) HY, hot-pressed yttria [17]; and (d) FY, fused yttria.

Name	Grain size (μm)	Young's modulus (GPa)	K_{IC} (GPa)
ASY	<2	146	3.27 ± 0.6
CY	1	153	2.28 ± 0.5
FY	0.8	—	—
HY [17]	>5	168	1.2 ± 0.5

Table 2.
Hardness, Young's modulus, and K_{IC} of yttria ceramics sintered.

and

weight loss rate (g/min cm^2) = weigh variation (g)/exposure area (cm^2) × exposure time (min) × weight before exposure (g) (3)

Quartz and Si wafer (100) were etched at a depth of 2.94–3.14 μm within 10 min of plasma irradiation, and then the etching evaluation was completed. On the other hand, the etching depth of Al_2O_3 and ZrO_2 irradiated with plasma for 60 min was 1.31–1.45 μm. The resistivity of the yttria was 0.41–0.54 μm, which was three times higher than that of the control.

The etching depth and etch rate of the control and yttria specimens was in the order of Si wafer, quartz, ZrO_2, Al_2O_3, and Y_2O_3, as shown in **Figure 10(a)**. The weight loss rate of Al_2O_3 and ZrO_2 was 12 times higher than that of Y_2O_3, as shown in **Figure 10(b)**. In the present study, there was no difference in plasma resistance between the specimens according to the starting materials.

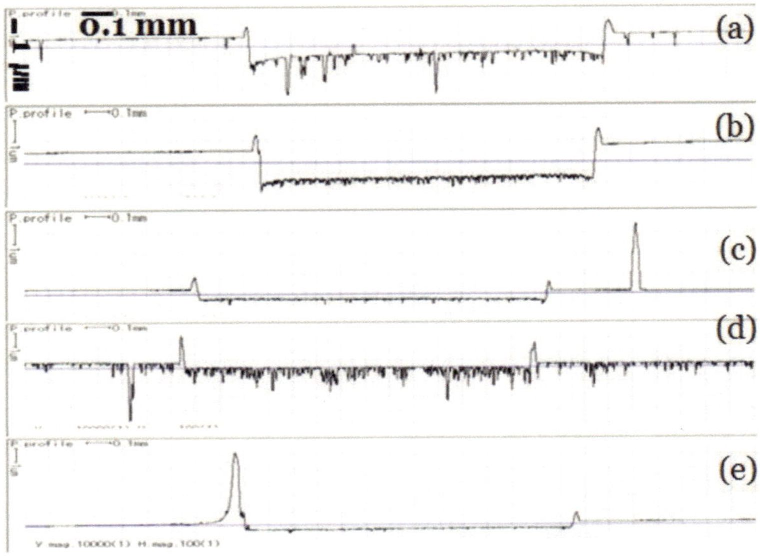

Figure 9.
Coverage of the etching profile for the topographies of the different samples by a plasma test: (a) Al_2O_3, (b) ZrO_2, (c) CY, (d) FY, and (e) HY [17].

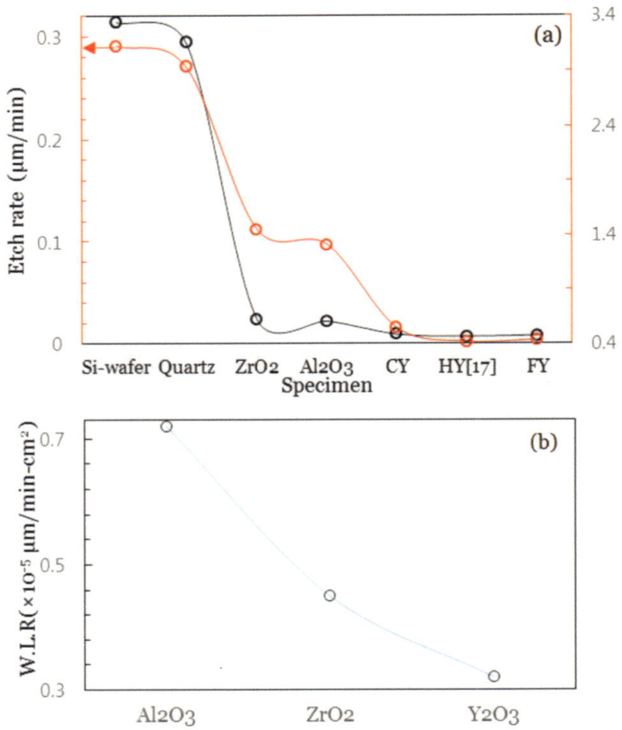

Figure 10.
Coverage of the plasma-deposited yttria ceramics versus quartz, silicon, Al_2O_3, and ZrO_2. (a) Etch depth and rate and (b) weight loss rate and relative weight loss rate. Deposition time of Si wafer (100) and quartz: 10 min. ZrO_2, Al_2O_3, and yttria (CY and HY [17]): 60 min. W. L. R. means the weight loss rate.

1.5 Summary

The plasma resistance and characteristics of the yttria ceramics were investigated in terms of calcination and three-step sintering. The crystal phase of the calcined yttria ceramics was identified as Y_2O_3, but as-received yttria ceramics contained both Y_2O_3 and $YO_{1.335}$ phase on the XRD pattern have been demonstrated.

The coarse particles in the initial stage changed to uniform particles after repeat-cycle sintering of yttria in air. In the repeated heat treatment process, the liquid phase with local differences in the Y:O ratio was converted to a uniform crystal grain.

The stress propagation along the cleavage plane and the low-level mist shape appeared to the hot-pressed specimen. The Young's modulus increased with decreasing particle size, and K_{IC} appeared to be relatively high regardless of the particle size because the dense microstructure, crystal phase, and the amorphous phase were mixed. The etching depth and etch rate of the control and yttria specimens were in the order of Si wafer, quartz, ZrO_2, Al_2O_3, and Y_2O_3. The weight loss rate of Al_2O_3 and ZrO_2 was 12 times higher than that of yttria ceramics. There was no difference in plasma resistance between the specimens according to the starting materials. The calcination and three-repeated cycle could be suggested as one of effective sintering methods of yttria.

2. Characteristics of thermal oxidation on hot-pressed pure yttria ceramics

2.1 Introduction

Fundamental studies on the forming process that affects the sintering behavior of yttria ceramics and powder synthesis techniques to obtain full density have attracted considerable interest. In particular, the sintering temperature has direct effects on the porosity, density, microstructure or crystal phase, and grain boundary migration. Yttria is used widely as a liquid sintering additive in AlN, SiC, SiAlON, and ZrO_2 [7, 12]. Yttria itself, however, is a sintering-limit material when heated and can induce abnormal grain growth during the sintering process [6].

Finely divided yttria, having a purity at least 99.9%, was pressed into compacts, sintered in a dry hydrogen atmosphere or a partial vacuum to 2150–2300°C, and refired in wet hydrogen at 1950–2300°C to redoxidize any yttria that had been reduced to yttrium during sintering [8]. The resulting polycrystalline yttria ceramics were examined to determine if there was any improvement in the intrinsic properties of yttria.

The previously studied yttria ceramics were divided into three groups. The first was developed from the control of yttria particles, such as 5 nm Y_2O_3 synthesized from a 50 to 120 nm Y_2O_3 powder and $Y(OH)_3$ using a coprecipitation method [9, 24]. The second group was a transparent yttria ceramic used as an infrared window to examine the electrical properties of yttria ceramics with additives, Th, Nd, and Er [15, 28, 29]. The last group involved the development of unconventional sintering techniques, such as triaxial isostatic press, hot isostatic pressing, or vacuum sintering, to fabricate full-density polycrystalline yttria ceramics [13, 30].

Most studies reported that the synthesis of nano-yttria particles was dependent only on the process variables. On the other hand, derivation of a common denominator for the properties of yttria is more complex because of the sintering methods of yttria ceramics, sintering property, and grain boundary migration by dopants such as La, Sr, and Sc [25, 31].

Nevertheless, the purity of the starting material and packing behavior as a function of the particle size used in the granule preparation together with the particle surface area affects the sintering characteristics [13]. These are important factors in the synthesis of dense sintered bodies. When yttria powders are induced on the nanoscale and have high purity, the fine particles or submicron-sized secondary particles agglomerate easily. Although deformation occurs in most agglomerated particles under applied stresses, strongly agglomerated secondary particles maintain their agglomerated shapes. Because the green compact containing secondary agglomerated particles produces discontinuous microstructures at the interface of the agglomerated zones to show a difference in heat transfer, its evasion is desirable.

From this point of view, a high-purity 99.9% Y_2O_3 powder as a starting material and a hot isostatic pressing method (hot-pressed), which are useful for obtaining a fully dense ceramic, were used to observe the sintering behavior as a function of temperature.

A foundation study was performed through comparisons between the microstructure of hot-pressed yttria ceramics as a function of temperature and that of the density, crystal phase, weight loss, Vickers hardness, the behavior of indentation, flexural strength, etc.

2.2 Characteristics of the hot-pressed yttria ceramics

Figure 11 shows the X-ray diffraction (XRD) patterns of hot-pressed yttria ceramics as a function of temperature. The sintering temperature in hot isostatic pressing (FRET-18, Fuji Denfa, Japan) was increased from 1300 to 1800°C at 100°C intervals, where the heating rate, holding time, and Ar gas flow were 5°C/min, 8 h, and 3 kgf/cm^2, respectively.

The hot isostatic pressing (hot-pressed) yttria ceramic specimens were oxidized at 1200°C for 12 h. XRD revealed the characteristic peaks of cubic yttria (JCPDS 41-1105), as shown in **Figure 11(a)** and **(b)**, irrespective of the sintering temperature and oxidation reaction at 1200°C.

In previous studies using a conventional sintering method, the crystallite phase of yttria could not be described precisely as the yttria phase because the $YO_{1.401}$ (JCPDS39-1064) and $YO_{1.335}$ (JCPDS 39-10,654) phases were also detected in the XRD pattern [25, 31]. This was based on the possibility that lattice distortion due to cation invasion into the oxygen vacancies or the Y-O ions can escape the regular lattice [32]. On the other hand, yttria ceramics containing Sc, Gd, La, Yb, and Mg were accompanied by a change in translucency and grain boundary migration, but the crystal phase was not changed [5].

The change in the intensity ratio reversed the peak at 48 and 57° in the high-temperature region, 1800°C, as shown in **Figure 11(b)**. Therefore, it was necessary to further examine the fraction of the crystal lattice due to the decomposition and volatilization of the yttria component [19, 31]. This means that the yttria crystalline phase varied according to the synthesis conditions and sintering methods, but the Y_2O_3 crystal phase identified in hot-pressed yttria ceramics means that the sintering temperature and oxidation reaction had no effect [18, 33].

Figure 12 shows the hot-pressed yttria ceramics sintered as a function of temperature. As the sintering temperature was increased, the color of the sintered body changed to black. This is dependent on the weight deviation and change in color due to an increase in oxygen defect concentration and Y:O ratio variation [12, 18]. This suggests that the black samples obtained from a traditional sintering method are due to the oxygen released in the lattice site during the sintering process.

In addition, the proposed hot-pressed yttria ceramics could be applied to marking samples with a crystalline structure and full density, because it has a larger effect on the color change than repeat-cycle sintering or conventional sintering. The oxidation reaction specimens were converted to white regardless of the sintering temperature, which suggests that oxygen diffusion affects the

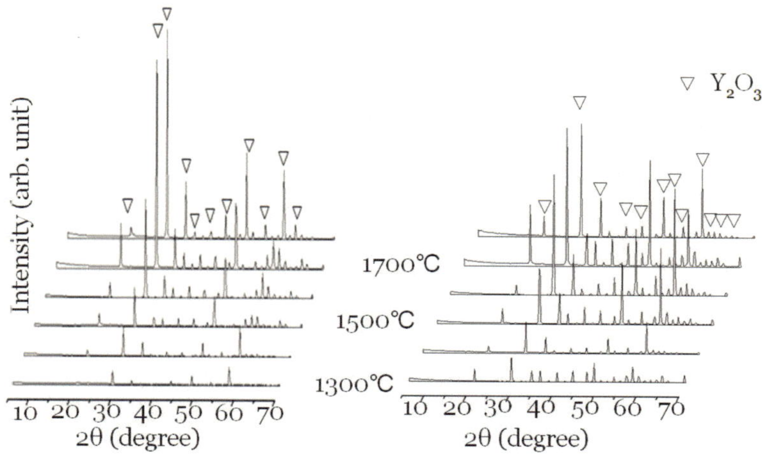

Figure 11.
XRD patterns of yttria ceramic. (a) Hot-pressed yttria ceramics as a function of temperature and (b) tempered in an oxidation of the hot-pressed yttria ceramics at 1200°C in air for 8 h.

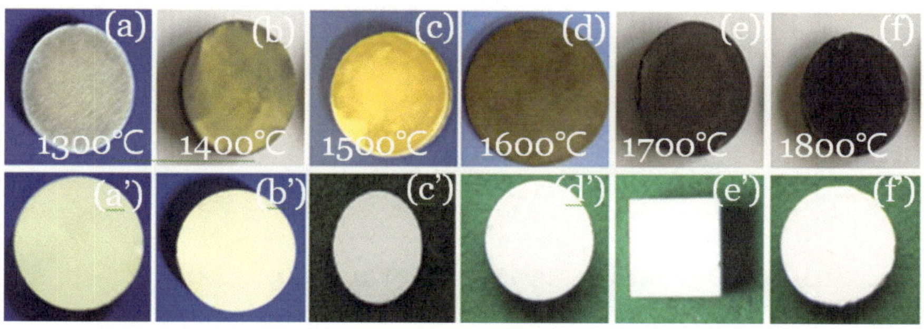

Figure 12.
Colors of yttria samples sintered. (a)~(f) Hot-pressed yttria ceramics as a function of temperature and (a')~(f') tempered in an oxidation of the hot-pressed yttria ceramics at 1200°C in air for 8 h.

oxygen vacancies in the lattice directly [21]. The 1500°C specimen was translucent and showed the considerable development of a glassy phase with increasing sintering temperature.

The weight of the specimens after the oxidation of yttria ceramics showed a tendency to increase with increasing temperature in **Figure 13**. The self-diffusion coefficient of yttrium, D_o, and activation energy, Q, for polycrystalline Y_2O_4 with a density of 99.9% at 1400–1700°C were reported to be 1.65×10^{-2} cm^2/s and 289 J/mole, respectively [31]. The D_o and Q values in the interstitial mechanism at 1050–1250°C were >6×10^{-6} cm^2/s and 82 kJ/mole, respectively [29]. The self-diffusion coefficient was expressed as $D = Do\ exp(-Q/RT)$. O had 3.5 times the activation energy of Y, and the self-diffusion of oxygen ions was relatively higher than yttrium. In addition, the diffusion of oxygen anions means that they occur relatively quickly because the interstitial diffusion of yttrium cations was similar to the rate of grain boundary migration [5, 18, 21]. Therefore, the diffusion rate during the oxidation of the hot-pressed yttria ceramics was higher than that of Y atom, and the oxygen vacancy showed preferential filling. This suggests that the weight increases with increasing weight of oxygen (see **Figure 14**).

This tendency corresponded to an approximate 35% reduction in weight due to gas evolution and volatilization in the precursor at temperatures up to 1250°C [10]. The density of the hot-pressed yttria ceramics of 1600°C was close to those of the theoretical density of yttria, 5.03 g/cm^3, and the density tended to decrease

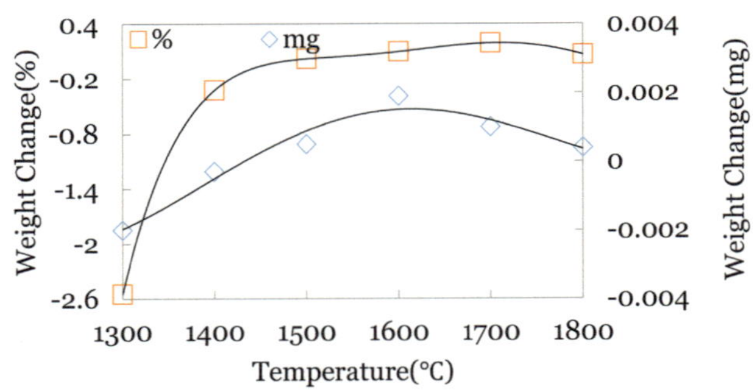

Figure 13.
Behavior of weight of hot-pressed yttria ceramics after oxidation at 1200°C for 12 h in air.

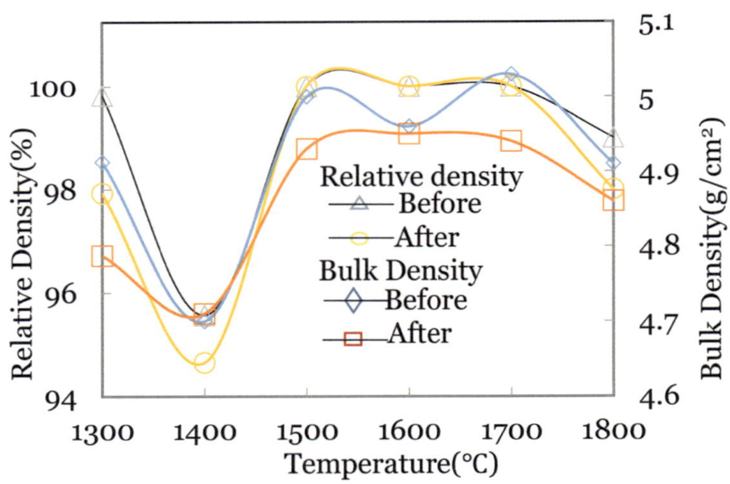

Figure 14.
Bulk and full density of hot-pressed yttria ceramics.

with increasing temperature. **Figure 15** presents SEM images of the microstructure according to the sintering temperature.

The initial 0.5 μm uniform crystal grains changed to a glassy phase, in which the grain size was unclear depending on the sintering temperature, as shown in **Figure 15(a)–(d)**. **Figure 15(e)** and **(f)** presents the grain shape of the pentagonal or hexagonal of 15 μm grain size in yttria ceramics hot pressed at 1700°C. Grain growth tended to show a continuous increase with increasing temperature, which has a direct effect on the porosity, density, and crystal phase (see **Figure 15(a′)–(f′)**). On the other hand, the microstructure of hot-pressed yttria ceramics had an obscurity grain boundary like a glassy phase after the sintering process, which was illustrated by the change in density and weight from **Figure 15(e′)** and **(f)**. The Young's modulus of 120–170 GPa measured by the pulse echo method using an oscilloscope with X and Y modulation in **Figure 16** is believed to be dependent on the yttria ceramic microstructure. **Figure 17** shows the behavior of Vickers indentation of the specimen sintered as a function of temperature.

The indentation shape of the 1400°C specimen showed slightly ragged edges and interior surface, whereas in the 1500°C specimen, the origin, mirror, and mist-like noncrystalline fracture surface were distinguished clearly. The relationship between the grain size and crystal phase was examined by measuring the shape and length of the indentation according to the sintering temperature. This tendency increased with increasing indentation crack length when the crystal

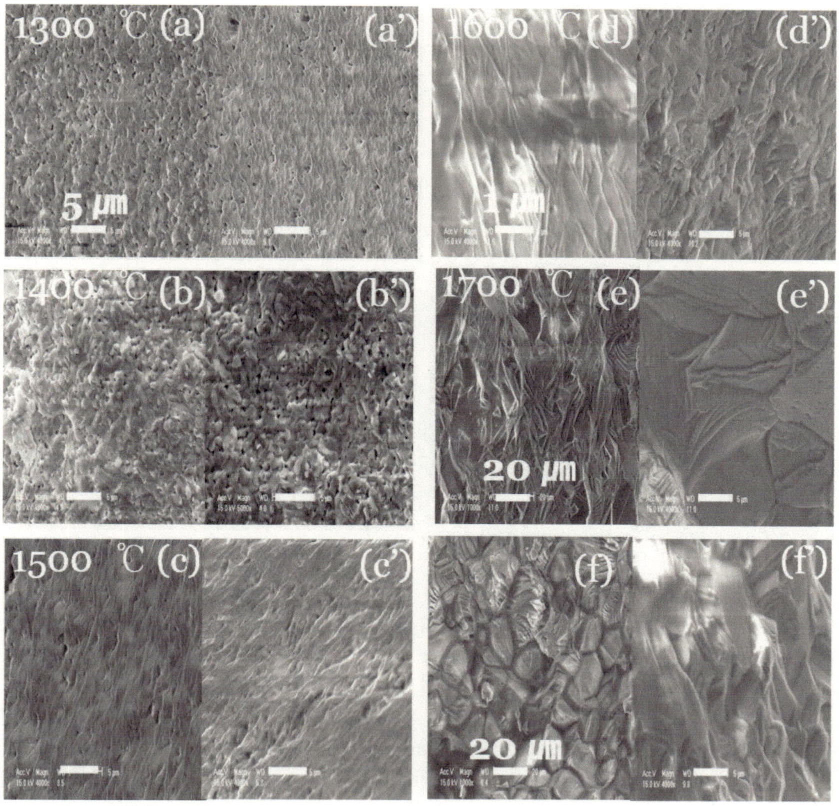

Figure 15.
SEM images of yttria sintered by hot pressing as a function of temperature. (a)~(f) Hot-pressed and (a')~(f') oxidized at 1200°C for 8 h in air.

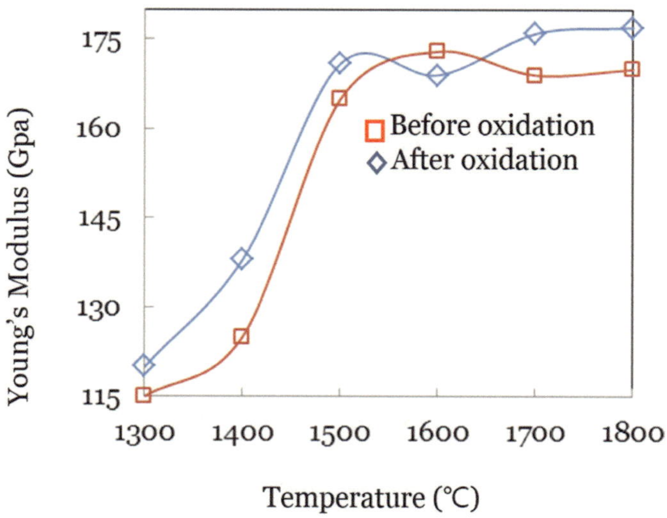

Figure 16.
Young's modulus of yttria ceramic hot-pressed as a function of temperature.

grains of the $MgAl_2O_4$ ceramics were grown from 0.4 to 24 μm and were attributed to the presence of local cracks acting as grain spalling [26]. **Figure 18** shows the K_{IC} and Vickers hardness measured by applying a load of 500 g according to JIS R1610 method. K_{IC} was calculated to be 1.2–1.9 MPa. The displacement point

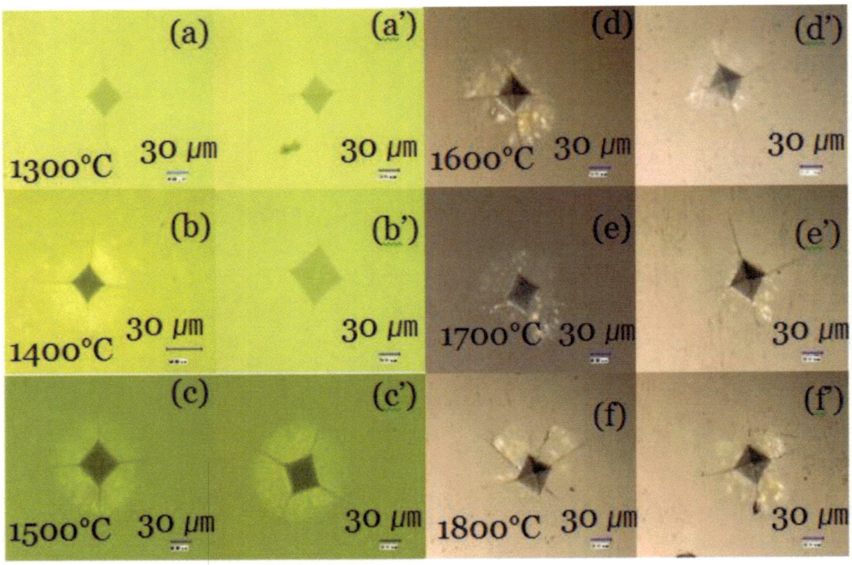

Figure 17.
Indentation images of hot-pressed yttria ceramics; (a)~(f) hot-pressed yttria ceramics as a function of temperature and (a')~(f') tempered in an oxidation of the hot-pressed yttria ceramics at 1200°C in air for 8 h.

of K_{IC} is dependent on the crystal phase according to the sintering temperature. The Vickers hardness of hot-pressed yttria ceramics was 28 GPa, indicating similar properties to those of the conventional polycrystalline yttria ceramics in **Figure 18(b)** [24].

The Vickers hardness of the hot-pressed yttria ceramics had high hardness because the particle size was smaller due to the Hall-Petch effect, but the hardness behavior tended to decrease with increasing particle size [27, 34]. The correlation between K_{IC} and the hardness in hot-pressed yttria ceramics was affected by the

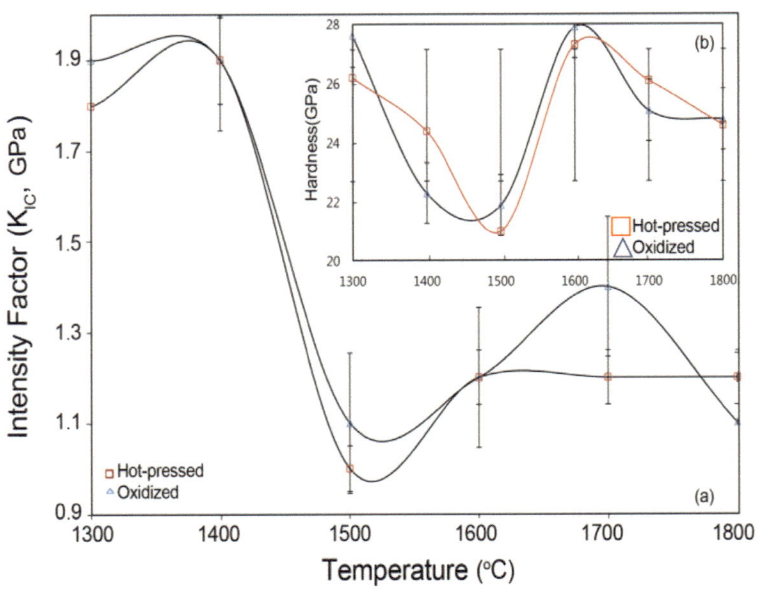

Figure 18.
Hardness and K_{IC} of hot-pressed yttria ceramic. (a) K_{IC} and (b) hardness, 500 g loads, giving typical Vickers indent diagonals. The vertical bars are the standard deviations.

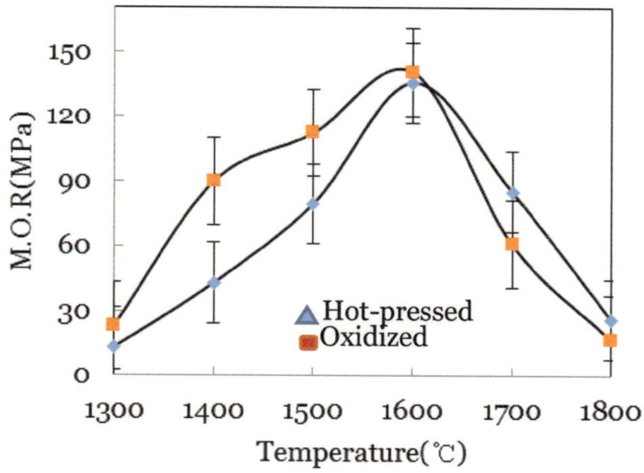

Figure 19.
Flexural strength of hot-pressed yttria ceramics.

grain size and the presence or absence of pores. The mean bending strength of five samples for each specimen was determined according to the JIS R1601 method. **Figure 19** shows the bending strength of the specimen. The specimens before heat treatment showed high strength in the high-temperature region, and those after heat treatment showed relatively high strength in the low-temperature region. The bending strength of the hot-pressed yttria ceramics was approximately 140 MPa, and it was similar to that of fused silica in **Table 3**.

Materials	Fracture toughness (MPa$\sqrt{m}$)	Strength (MPa)
Zirconia (3 mol% Y_2O_3)	7.0–12.0	800–1500
Silicon carbide		
Hot-pressed	4.8–6.0	230–825
Sintered	14.8	96–520
Silicon nitride		
Hot-pressed	4.1–6.0	700–1000
Reaction bonded	3.6	250–345
Sintered	5.3	414–650
Alumina oxide		
99.9% pure	4.2–5.9	282–551
96%	3.85–3.95	358
Hot-pressed yttria	1.0–1.9	20–140
Glass ceramics (Pyroceram)	1.6–2.1	123–370
Fused silica	0.79	104
Borosilicate glass (Pyrex)	0.77	69
Soda-lime glass	0.75	69
Polyethylene terephthalate (PET)	5.0	59.3
Polypropylene (PP)	3.0–4.5	31.0–37.2

Source: ASTM Handbooks, Vol. 1 and 19, Engineered Materials Handbook, Vol. 2 and 4, and Advances Materials & Processes, Vol. 137, No. 6 ASM International Materials Park, OH.

Table 3.
Fracture toughness and strength of various materials.

2.3 Summary

This study examined the characteristics of hot-pressed pure yttria ceramic that was annealed in an oxidation atmosphere. As the sintering temperature was increased, the color of the sintered body changed to black. The weight of the specimens after the oxidation of yttria ceramics showed a tendency to increase with increasing temperature. The hot-pressed yttria ceramics could be applied to marking samples with a crystalline structure and full density. Grain growth tended to show a continuous increase with increasing temperature, which has a direct effect on the porosity, density, and crystal phase. The relationship between the grain size and crystal phase increased with increasing indentation crack length and was attributed to the presence of local cracks acting as grain spalling. K_{IC} was calculated to be 1.2–1.9 MPa, and the Vickers hardness of hot-pressed yttria ceramics was 28 GPa. The specimens before heat treatment showed high strength in the high-temperature region, and those after heat treatment showed relatively high strength in the low-temperature region. The bending strength of the hot-pressed yttria ceramics was approximately 140 MPa. The yttria crystalline phase varied according to the synthesis conditions and sintering methods, but the Y_2O_3 crystal phase identified in hot-pressed yttria ceramics means that the sintering temperature and oxidation reaction had no effect.

Author details

Jin Sam Choi
School of Materials and Engineering, University of Ulsan, Ulsan, R.O. Korea

*Address all correspondence to: jinsamchoi03@gmail.com

References

[1] Hönninger C, Zhang G, Keller U, Giesen A. Femtosecond Yb-YAG laser using semiconductor saturable absorbers. Optics Letters. 1995;**20**(23):2402-2024

[2] Takaichi K, Yagi H, Lu J, Shirakawa A, Ueda K, Yanagitani T, et al. Yb^{3+}-doped $Y_3Al_5O_{12}$ ceramics—A new solid-state laser material. Physica Status Solidi. 2003;**200**(1):5-7

[3] Kodo M, Soga K, Yoshida H, Yamamoto T. Low temperature sintering of polycrystalline yttria by transition metal ion doping. Journal of the Ceramic Society of Japan. 2009;**117**(6):765-768

[4] Harris DC. Materials for Infrared Windows and Domes. Washington, USA: SPIE-The International Society for Optical Engineering; 1999

[5] Lapato LM, Shevchenko AV, Kushchevskii AE. Polymorphic transitions of rare earth oxides at high temperatures. Inorganic Materials. 1974;**10**(8):1276-1281

[6] Carlsson ON. The O-Y (oxygen-yttrium) system. Bulletin of Alloy Phase Diagrams. 1990;**11**(1):61-66

[7] Iwasawa J, Nishimizu R, Tokita M, Kiyohara M, Uematsu K. Plasma resistance dense yttrium oxide film prepared by aerosol deposition process. Journal of the American Ceramic Society. 2007;**90**(8):2327-2332

[8] Muta A, Tsukuda Y. Method for Sintering Very Pure Yttria Compacts to Transparency. U.S. Pat. 3,764,643; 1973

[9] Chen IW, Wang XH. Sintering dense nanocrystalline ceramics without final-stage grain growth. Nature. 2000;**404**:168-171

[10] Saito N, Matsusa SI, Ikegami T. Fabrication of transparent yttria ceramics at low temperature using carbonate-derived powder. Journal of the American Ceramic Society. 1998;**81**(8):2023-2028

[11] Chen PL, Chen IW. Grain boundary mobility in Y_2O_3: Defect mechanism and dopant effects. Journal of the American Ceramic Society. 1996;**79**(7):1801-1809

[12] Jorgensen PJ, Anderson RC. Grain-boundary segregation and final stage sintering of Y_2O_3. Journal of the American Ceramic Society. 1967;**50**(11):553-558

[13] Huang Y, Jiang D, Zhang J, Lin Q. Fabrication of transparent lanthanum-doped yttria ceramics by combination of two-step sintering and vacuum sintering. Journal of the American Ceramic Society. 2009;**92**(12):2883-2887

[14] Druschitz AP, Schroth JG. Hot isostatic pressing of a presintered yttria-stablized zirconia ceramic. Journal of the American Ceramic Society. 1989;**72**(9):1591-1597

[15] Greskovich C, Woods KN. Fabrication of transparent ThO_2-doped Y_2O_3. Journal of the American Ceramic Society Bulletin. 1973;**52**(5):47378

[16] Katayama K, Osawa H, Akiba T, Yanagida H. Sintering and electrical properties of CaO-doped Y_2O_3. Journal of the European Ceramic Society. 1990;**6**:39-45

[17] Choi JS, Shin DW, Bae WT. Characteristics of thermal oxidation on hot-pressed pure yttria ceramics. Journal of the Korean Ceramic Society. 2013;**50**(3):180-185

[18] Chen PL, Chen IW. The role of defect interaction in boundary mobility and cation diffusivity of CeO_2. Journal of the American Ceramic Society. 1994;77(9):2289-2297

[19] Jollet F, Noguera C, Gautier M, Thromat N, Duraud JP. Influence of oxygen vacancies on the electronic structure of yttrium oxide. Journal of the American Ceramic Society. 1991;**74**(2):358-364

[20] Shannon RD. Revised effective ionic radii and systematic studies of interatomic distances in halides and chalcogenides. Acta Crystallographica. 1976;**A32**:751-767

[21] Kingery W, Bowen H, Uhlmann D. Introduction to Ceramics. 2nd ed. New York, U.S.A.: John Wiley & Sons Inc.; 1976. p. 418, 431 and 505

[22] Ikegami T, Li JG, Mori T. Fabrication of transparent yttria ceramics by low temperature synthesis of yttrium hydroxide. Journal of the American Ceramic Society. 2002;**85**(7):1725-1729

[23] Rhodes WH, Trickett EQ, Sordelet DJ. Key powder characteristics in sintered optical ceramics. Ceramic Transactions. 1990;**12**:677-690

[24] Nobby T, Kofstad P. Electrical conductivity and defect structure of Y_2O_3 as a function of water vapor pressure. Journal of the American Ceramic Society. 1984;**67**(12):786-792

[25] Swamy V, Dubrovinskaya NA, Dubrovinsky LS. High-temperature powder X-ray diffraction of yttria to melting point. Journal of Materials Research. 1999;**14**(2):456-459

[26] Rice RW, Carl CW, Borchelt F. Hardness-grain-size relations in ceramics. Journal of the American Ceramic Society. 1994;77(110):2539-2553

[27] Armstrong RW, Raymond EL, Vandervoort RR. Anomalous increases in hardness with increase in grain size of beryllia. Journal of the American Ceramic Society. 1970;**53**:529-530

[28] Ikesue A, Kamata K, Yoshida K. Synthesis of transparent Nd-doped HfO_2-Y_2O_3 ceramics using HIP. Journal of the American Ceramic Society. 1996;**79**(2):359-364

[29] Berard MF, Wilder CD. Cation self-diffusion in polycrystalline Y_2O_3 and Er_2O_3. Journal of the American Ceramic Society. 1969;**52**(2):85-88

[30] Choi JS, Bae WT. Full Density Yttria Ceramic Sintered by Using for Conventional Sintering Method with Fused Yttria as Starting Materials. Kr. Pat. No: 10-2012-0082809; 2012

[31] Berard MF, Wirkus CD, Wilder DR. Diffusion of oxygen in selected rare earth oxides. Journal of the American Ceramic Society. 1968;**51**(11):643-647

[32] Ackermann R, Rauh E, Walters R. Thermodynamics study of the system yttrium + yttrium sesquioxide A refinement of the vapor pressure of yttrium. The Journal of Chemical Thermodynamics. 1970;**2**:139-149

[33] Verkerk MJ, Winnubst AJA, Burggraf AJ. Effect of impurities on sintering and conductivity of yttria-stabilized zirconia. Journal of Materials Science. 1982;**7**:3112-3113

[34] Christopher S, Nieh TG. Hardness and abrasion resistance of nano crystalline nickel alloys near the Hall Petch breakdown regime. Materials Research Society Symposium Proceedings. 2003;**740**:27-32

Chapter 5

Impedance Spectroscopy: A Powerful Technique for Study of Electronic Ceramics

Shukdev Pandey, Devendra Kumar, Om Parkash and Lakshman Pandey

Abstract

Electronic ceramics are technological materials having a vast variety of applications such as actuators and sensors, computer memories, electrically controlled microwave tuning devices for RADAR, etc. and are playing key role in electronics industry today. An electronic ceramic component can be visualised as grain-grain boundary-electrode system. Impedance spectroscopy is being widely used to separate out contributions of these to the overall property of a ceramic. This involves equivalent circuit models. To facilitate development of suitable equivalent circuit models and obtain values of the components, some most useful circuits with their simulated behaviour are presented. Steps highly useful in the modelling process are summarised. The procedure of impedance spectroscopy is illustrated by analysing the impedance data of the ceramic system $BaFe_xTi_{1-x}O_3$ (x = 0.05) containing two phases.

Keywords: electronic ceramics, impedance spectroscopy, equivalent circuit models, CPE, CNLS, grain-boundaries, ceramic-electrode interface

1. Introduction

Ceramics are inorganic non-metallic solids that have been processed and shaped by heating at high temperatures. Modern ceramics include oxides, nitrides, carbides, etc. and constitute a large fraction of technologically useful materials at present. Ceramics, whose electrical, magnetic, or optical properties are used in devices are broadly termed as electro-ceramics and are being heavily used as electrical insulators, TV baluns, mobile antennas and speakers, substrates for electronic circuits, computer memories, magnetic recording heads, high-temperature heating elements, cryogenic sensors, microwave tuning devices for RADAR applications, etc. [1–3]. The as-prepared ceramics are usually in powder form that are processed and shaped for device applications by subjecting to suitable sintering procedure. A ceramic piece thus consists of small crystallites called grains that are joined together in random orientations. The joining region called the grain boundary has, due to mismatch, strained bonds. Properties of grain boundary, therefore, are different from those of grains and highly depend upon the processing variables such as heating/cooling rates, presence of external fields, atmosphere, etc. By changing these variables and starting ingredients, behaviour of grains and grain boundary may be altered. It is the interplay between the grain and grain boundary behaviour that bestows ceramics with some very useful properties. Ceramics prepared by controlled crystallisation of glass are called glass-ceramics where the grain boundary consists of the uncrystallised glass [1, 4, 5]. The understanding and controlling of this interplay with the help of processing variables, additives, ingredients, reduction in grain size dwelling in nanometre range or compositing, with a view to develop desired, unforeseen and technologically useful properties, is the subject of intensive research activities at present. For device applications, a ceramic piece is usually connected to some other system through electrodes. Thus, an electronic component may be treated as grain-grain boundary-electrode system where the overall observed properties would have contributions from all these [1]. In order to develop ceramics having desired and reproducible properties, these individual contributions must be separated out. For this, impedance spectroscopy is a very useful technique [6]. It not only helps in developing the understanding of the processes but also provides with an equivalent circuit model for the system that may be used for simulation purposes [1, 6]. Impedance spectroscopic studies proceed with the measurement of electrical impedance as function of frequency and interpretation of the data by using suitable equivalent circuit model that seems appropriate to represent various charge transfer processes in the system. The values of the components used in the model are then obtained by using a least-squares procedure to fit the experimental impedance data with that simulated by using the chosen model. The choice of the most suitable model is a difficult process. Even if a model has been conceived, estimating the values of the components of the model, so much required for the least-squares programs, is not always straightforward. This poses some inconvenience to the worker. Developing a suitable model and obtaining these estimates are greatly facilitated by comparing the experimental plots with those simulated for various models and considering different charge transfer processes thought to be possibly present in the system. In what follows, basics of impedance spectroscopy and some very useful models with their simulated behaviour are presented. Analysis of impedance spectroscopic data on ceramic system $BaFe_xTi_{1-x}O_3$ ($x = 0.05$) containing two phases is also presented as illustration. The focus is how to choose a model and how to get an estimate of the values of the components used in the model.

2. Basics of impedance spectroscopy

Impedance spectroscopy essentially involves measurement of real and imaginary parts of electrical impedance ($Z^* = Z' - j Z''$, $j = \sqrt{(-1)}$) of a system as function of frequency ($\omega = 2\pi f$) for various parameters of interest such as composition, temperature, etc. The values of Z' and Z'' are plotted as function of frequency (Z', Z'' vs. log f) and also in complex plane (i.e. Z'' vs. Z'). These complex plane plots are usually distorted semicircular overlapping arcs. By looking at the shapes of these plots, charge transfer processes present in the system and accessible in the measurement frequency range are inferred. This is greatly facilitated by comparing the experimental plots with those simulated for possible equivalent circuit models for the system. Detailed discussion on impedance spectroscopy and equivalent circuit models has been presented by Barsoukov and Macdonald [6], Jonscher [7], and Pandey et al. [8]. Due to ready availability of versatile impedance analysers working in extended frequency ranges and ease of measurements, impedance spectroscopy has emerged as a very popular and powerful tool in recent years and is being widely used in various fields encompassing materials technology, electrochemistry, biology, medical diagnostics, agriculture, dairy and fruit production ([9] and references therein).

The electrical behaviour of a system can be expressed in terms of interrelated functions known as impedance ($Z^* = Z' - j Z''$), admittance ($Y^* = (Z^*)^{-1} = Y' + j Y''$), permittivity ($\varepsilon^* = (j \omega C_0 Z^*)^{-1} = \varepsilon' - j \varepsilon''$), and modulus ($M^* = (\varepsilon^*)^{-1} = j\omega C_0 Z^* = M' + j M''$), C_0 being the capacitance of the empty cell used to house the sample. These are broadly termed as immittance functions and are conveniently used to develop equivalent circuit models for the sample-electrode system [6, 8, 9]. A charge transfer process would have a certain time constant and would respond in the corresponding frequency region. A parallel RC circuit (**Figure 1(a)**) possesses one time constant RC and is thus conveniently used to represent one charge-transfer process. The impedance of this model circuit is given as

$$Z' = \frac{R}{1 + (\omega CR)^2}, \quad Z'' = \frac{\omega CR^2}{1 + (\omega CR)^2} \tag{1}$$

Z' and Z'' satisfy the relation

$$\left(Z' - \frac{R}{2}\right)^2 + Z''^2 = \left(\frac{R}{2}\right)^2 \tag{2}$$

which is equation of a circle having centre at point (R/2, 0), radius equal to R/2, intercept with Z' axis at the point (R,0), and passing through origin in Z'' vs. Z' plot. As the values of R and C are positive, the Z'' vs. Z' plot will be a semicircle (**Figure 1(b)**). On this semicircle, the peak point occurs at $Z' = R/2$. At this point, Z'' is equal to R/2 and $\omega CR = 1$. Also, $Z' = R$ at $\omega = 0$. Thus, if values Z' and Z'' for a sample are experimentally measured in certain frequency range and Z'' vs. Z' plot is found to be a clear semicircular arc, then a parallel RC circuit model may be used to represent the electrical behaviour of the sample, to begin with (for more details see Section 3.1). The value of intercept of the extrapolated arc with the Z' axis towards

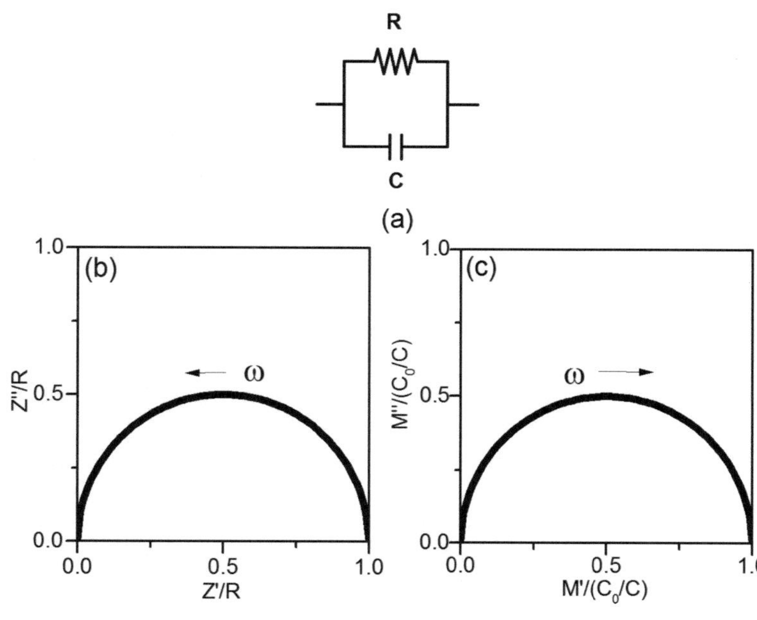

Figure 1.
(a) Equivalent circuit model containing parallel combination of R and C. Normalized (b) Z''/R vs Z'/R and (c) M''/(C₀/C) vs M'/(C₀/C) plots.

low-frequency side will give the value of R. The value of C can be obtained by noting the frequency where the arc peaks and using the relation $\omega CR = 1$.

For this model, M′ and M″ are given as

$$M' = \omega C_0 \left(\frac{\omega CR^2}{1 + \omega^2 C^2 R^2} \right), M'' = \omega C_0 \left(\frac{R}{1 + \omega^2 C^2 R^2} \right) \quad (3)$$

which satisfy the relation

$$\left(M' - \frac{C_0}{2\,C} \right)^2 + M''^2 = \left(\frac{C_0}{2C} \right)^2 \quad (4)$$

Eq. (4) indicates that M″ vs. M′ plot would be a semicircular arc passing through the origin and having intercept at $M' = C_0/C$ and centre at $(C_0/2C, 0)$ (**Figure 1(c)**). On this plot, M″ would peak at frequency where $\omega CR = 1$. Value of M′ becomes equal to C_0/C as $\omega \to \infty$. Thus, by using M″ vs. M′ plot of the same experimental data, C can be obtained by noting the high-frequency intercept in the M″ vs. M′ plot. The value of R can be obtained by using this value of C and noting the frequency where M″ peaks. In the M″ vs. M′ plot, the arc traverses from left to right whereas in Z″ vs. Z′ plot, it traverses from right to left as the frequency is increased.

Choice of a model to represent a system is a difficult process, becoming more so since same behaviour can be simulated by different models [6]. The choice is greatly facilitated by comparing the experimental plots with those simulated for various models and considering the processes that might be present in the system with a preference to simple models to start with. When the complex plane plots display more than one arcs, presence of more polarisation/charge transfer processes is inferred. A general practice is to consider one RC circuit to represent one process and connect more RCs in series to develop an equivalent circuit model. Thus for representing grain-grain boundary-electrode system in a ceramic, a model comprising three parallel RCs connected in series might be taken [1]. It has been found that usually combinations of resistances (R) and capacitances (C) suffice for dielectrics, combinations of R and inductance L suffice for magnetic systems, and combinations of R, L and C suffice for ferro-/piezoelectrics [6–8, 10–20]. Sometimes, it is found that the lumped-component type of models do not yield good fits and their simulated patterns do not show even qualitative resemblance with the experimental plots. In those cases, attempt is made to represent the data by models involving constant phase angle elements (CPE) [6, 9] that are considered to correspond to some sort of distribution in the material properties. The values of the components used in the model are estimated by comparing the experimental plots with the simulated ones. These values are obtained more accurately by fitting the experimental data with complex non-linear least-square (CNLS) procedure using these values as initial guesses. For cases, where it is not possible to decide upon an appropriate model, equivalent circuits models that seem to be most probable to represent the processes thought to be possibly present, or to be dormant but becoming dominant as some variables change, in the system may be considered and the model that yields the lowest value of sum of squares of errors in the CNLS runs may be accepted [6].

In order to facilitate a prompt development of an equivalent circuit model for electronic ceramics, some very important clues have been given in [9] and are summarised below for a ready reference.

1. The experimental data can be represented in any of the formalisms (Z, M, Y and ε) and interpreted. However, analysis by using only one formalism might lead to erroneous conclusions and two or more functions such as Z and M should be used [6, 8].

2. If plots of both Z″ vs. Z′ and M″ vs. M′ are clear semicircles passing through origin, then the system may be represented by a model having one parallel RC and presence of one process may be inferred [6, 8].

3. Appearance of a shift towards right in M″ vs. M′ plot (here, the arc traverses from left low-frequency side to right high-frequency side) and steeply rising branch at low-frequency side in the Z″ vs. Z′ plot (here, the arc traverses from right to left) indicates the presence of series capacitance C in the equivalent circuit [17].

4. Appearance of a shift towards right in Z″ vs. Z′ plot and steeply rising branch at high-frequency side in M″ vs. M′ plot indicates the presence of series resistance R in the equivalent circuit [8, 12].

5. Appearance of two clear semicircular arcs in Z″ vs. Z′ plot or M″ vs. M′ plot indicates presence of two processes and equivalent circuit would contain two parallel RC's (R_1C_1 and R_2C_2). A depressed looking semicircular arc in the Z or M plots would indicate the presence of at least two processes having ratio of time constants R_2C_2/R_1C_1 in the range 1–5 [8].

6. If the experimental Z″ vs. Z′ plot shows a semicircular arc with changed sign of Z″ in the whole frequency range (i.e. arc appears in fourth quadrant traversing from left to right), then a model with parallel R-L circuit may be used [11].

7. A cross over from positive values of Z″ to negative values or vice versa within the overall frequency range covered in the experiment indicates a situation of resonance and inclusion of R, C and L in the model would be needed [6, 14, 16].

8. Presence of a linear portion in the immittance plots would indicate the presence of series CPE in the model. Depressed looking immittance plots indicate that CPE connected in parallel may be included in the model [6, 9, 12, 21].

Detailed analysis of various models involving CPE is available in [6, 9]. It is believed that CPE represents distribution in properties. Therefore, when the data is well represented by a model involving CPE, presence of distribution in certain properties of ceramics is inferred. It may be mentioned that capacitor C and inductor L are used to represent the storage of electrical and magnetic forms of energy in a system. At frequencies much below resonance, the situation may be approximately modelled by considering RC's only. Similarly, a mechanically vibrating system can be analogously represented by an RLC circuit. Therefore, equivalent circuits having combinations of R, L and C may be used [14, 16] for piezoelectric ceramics. In what follows, some model circuits found very useful in study of ceramics are briefly presented. Emphasis is given to Z″ vs. Z′ and M″ vs. M′ plots as these were found more informative.

3. Equivalent circuit models

3.1 Model 1: parallel combinations R_1C_1 and R_2C_2 connected in series

Parallel combinations R_1C_1 and R_2C_2 connected in series [8] are shown in **Figure 2(a)** and the values for Z′ and Z″ are given in Eq. (5):

$$Z' = \frac{R_1}{1+(\omega C_1 R_1)^2} + \frac{R_2}{1+(\omega C_2 R_2)^2},\; Z'' = \frac{\omega C_1 R_1^2}{1+(\omega C_1 R_1)^2} + \frac{\omega C_2 R_2^2}{1+(\omega C_2 R_2)^2} \tag{5}$$

Complex plane plots for various immittance functions are shown in **Figure 2(b-e)** for $R_2/R_1 = 1$. It is seen that when the values of time constants are widely separated ($R_2C_2/R_1C_1 = 100$), two clear arcs appear indicating the presence of two processes. For $R_2C_2/R_1C_1 > 1$ but below 5, a depressed looking arc appears and when $R_2C_2/R_1C_1 = 1$, a single arc appears (**Figure 2(b)**).

Now let us look at the plots of **Figure 3(b-e)** for $R_2/R_1 = 100$. Here, Z'' vs. Z' plot obtained for different ratios of time constants looks like a single semicircular arc indicating presence of one process. However, the corresponding M'' vs. M' (**Figure 3(c)**) has two clear arcs indicating the presence of two processes. This indicates that inference derived from only one formalism might lead to wrong conclusions, and plots using more formalisms should be looked at together.

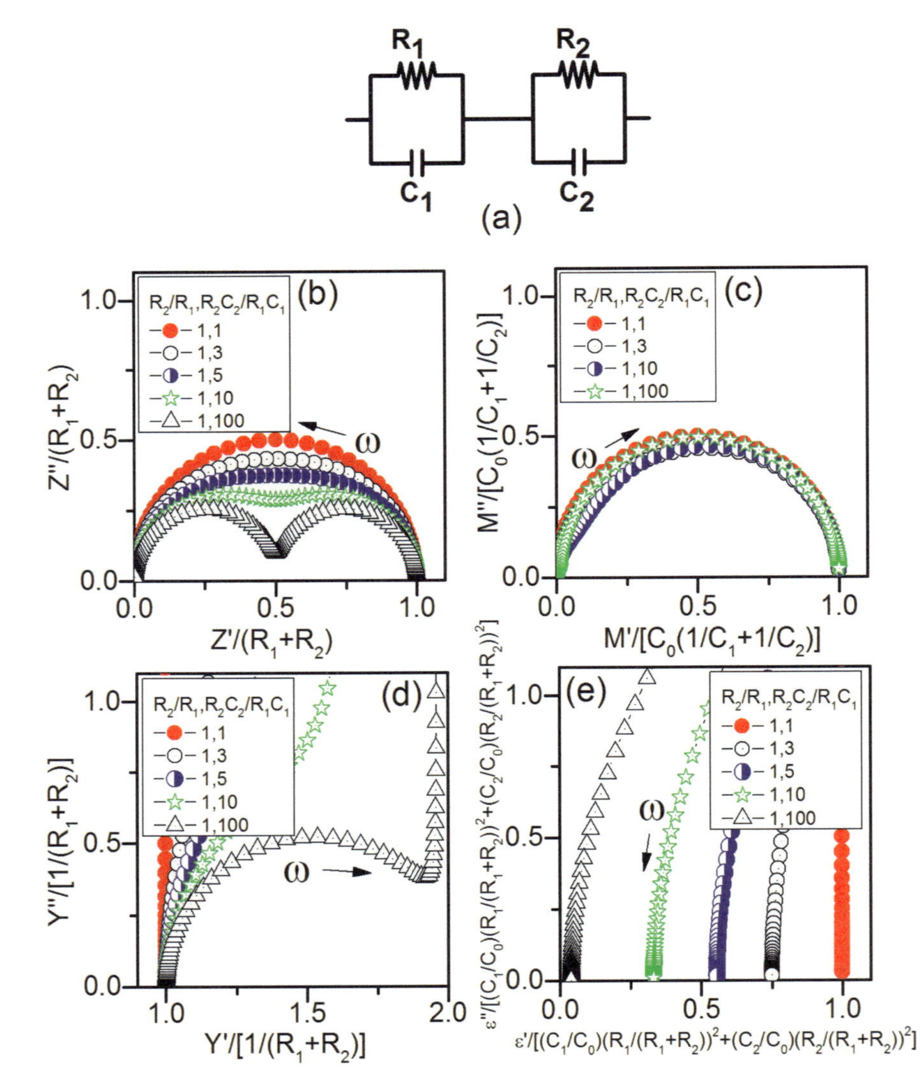

Figure 2.
(a) Equivalent circuit model containing series combination parallel R_1C_1 and R_2C_2. Normalized (b) $Z''(R_1 + R_2)$ vs $Z'(R_1 + R_2)$, (c) $M''/[C_0(1/C_1 + 1/C_2)]$ vs $M'/[C_0(1/C_1 + 1/C_2)]$, (d) $Y''/[1/(R_1 + R_2)]$ vs $Y'[1/(R_1 + R_2)]$ and (e) $\varepsilon''/[(C_1/C_0)(R_1/(R_1 + R_2))^2 + (C_2/C_0)(R_2/(R_1 + R_2))^2]$ vs $\varepsilon'/[(C_1/C_0)(R_1/(R_1 + R_2))^2 + (C_2/C_0)(R_2/(R_1 + R_2))^2]$ for $R_2/R_1 = 1$ and R_2C_2/R_1C_1.

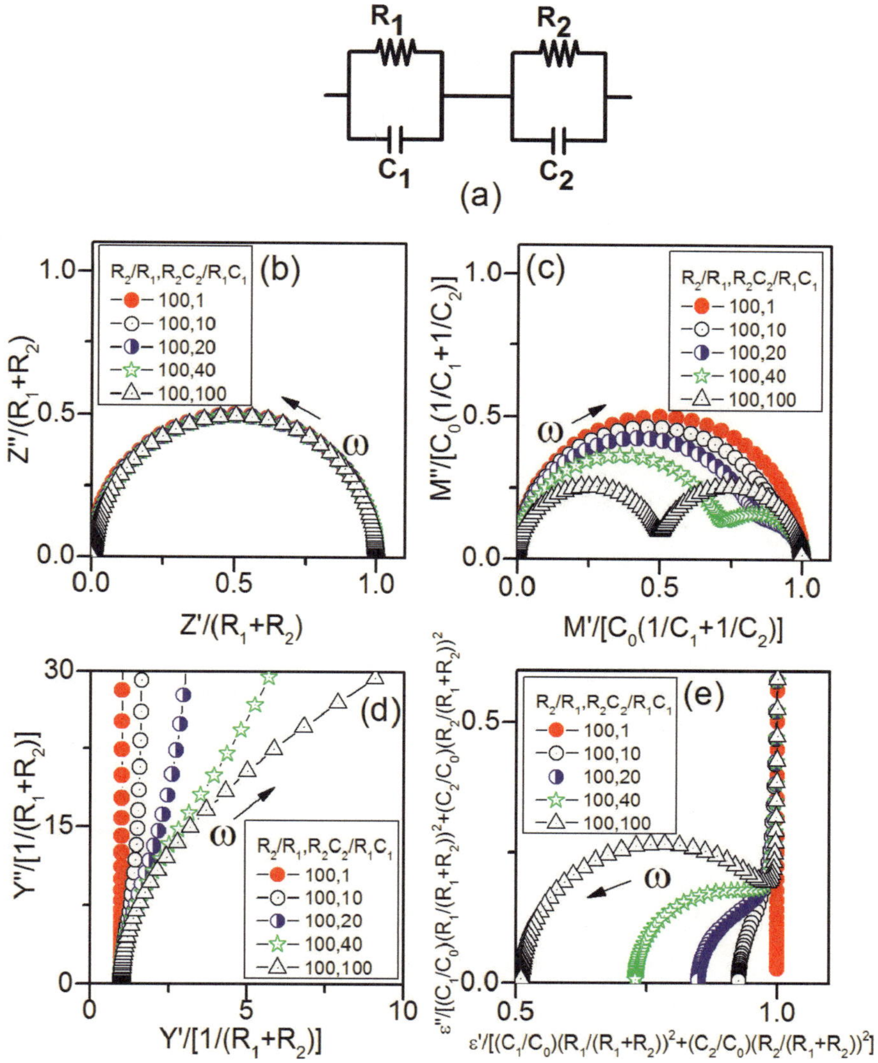

Figure 3.
(a) Equivalent circuit model containing series combination of parallel R_1C_1 and R_2C_2. Normalized (b) $Z''(R_1 + R_2)$ vs $Z'(R_1 + R_2)$, (c) $M''/[C_0(1/C_1 + 1/C_2)]$ vs $M'/[C_0(1/C_1 + 1/C_2)]$, (d) $Y''/[1/(R_1 + R_2)]$ vs $Y'[1/(R_1 + R_2)]$ and (e) $\varepsilon''/[(C_1/C_0)(R_1/(R_1 + R_2))^2 + (C_2/C_0)(R_2/(R_1 + R_2))^2]$ vs $\varepsilon'/[(C_1/C_0)(R_1/(R_1 + R_2))^2 + (C_2/C_0)(R_2/(R_1 + R_2))^2]$ for $R_2/R_1 = 100$ and different values of R_2C_2/R_1C_1.

3.2 Model 2: parallel combinations R_1C_1 connected in series with R_2

The model is schematically shown in **Figure 4(a)**. The values of Z′ and Z″ are given as

$$Z' = \frac{R_1}{1 + (\omega C_1 R_1)^2} + R_2 \quad , \quad Z' = \frac{\omega C_1 R_1^2}{1 + (\omega C_1 R_1)^2} \tag{6}$$

The immittance plots for this model are shown in **Figure 4(b-e)** for various values of R_2/R_1. It is seen that Z″ vs. Z′ plot (**Figure 4(b)**) shows a shift towards right, and the corresponding M″ vs. M′ plot (**Figure 4(c)**) has steeply rising branch at high-frequency side.

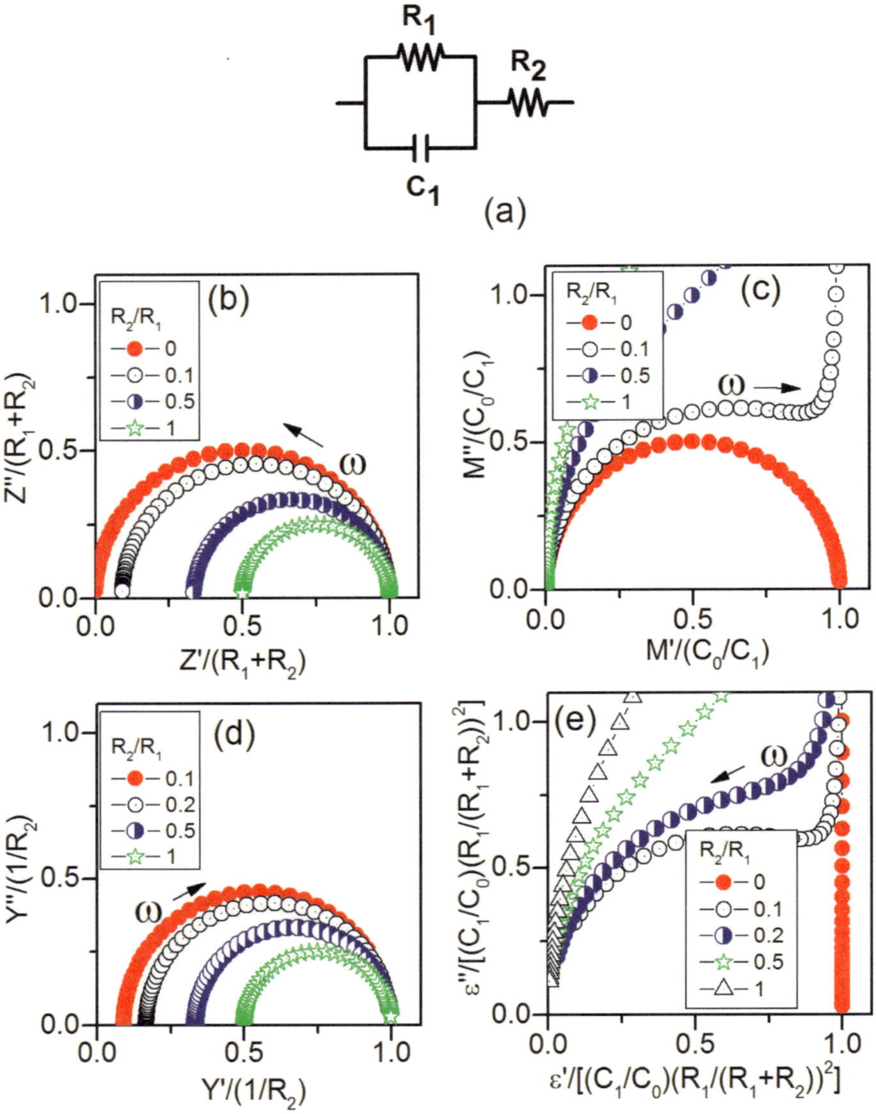

Figure 4.
(a) Equivalent circuit model containing series combination of parallel R_1C_1 and R_2. Normalized (b) $Z''/(R_1 + R_2)$ vs $Z'/(R_1 + R_2)$, (c) $M''/(C_0/C_1)$ vs $M'/(C_0/C_1)$, (d) $Y''/(1/R_2)$ vs $Y'/(1/R_2)$ and (e) $\varepsilon''/[(C_1/C_0)(R_1/(R_1 + R_2))^2]$ vs $\varepsilon'/[(C_1/C_0)(R_1/(R_1 + R_2))^2]$ for R_2/R_1.

3.3 Model 3: parallel combinations R_1C_1 connected in series with C_2

Parallel combinations R_1C_1 connected in series with C_2 is shown in **Figure 5(a)**. The values of Z' and Z'' are given as [17]

$$Z' = \frac{R_1}{1 + (\omega C_1 R_1)^2} \quad , \quad Z' = \frac{\omega C_1 R_1^2}{1 + (\omega C_1 R_1)^2} + \frac{1}{\omega C_2} \tag{7}$$

The immittance plots for this model are shown in **Figure 5(b-e)**. It is found that Z'' vs. Z' has a low-frequency rising branch where as a shift is seen in M'' vs. M' plot.

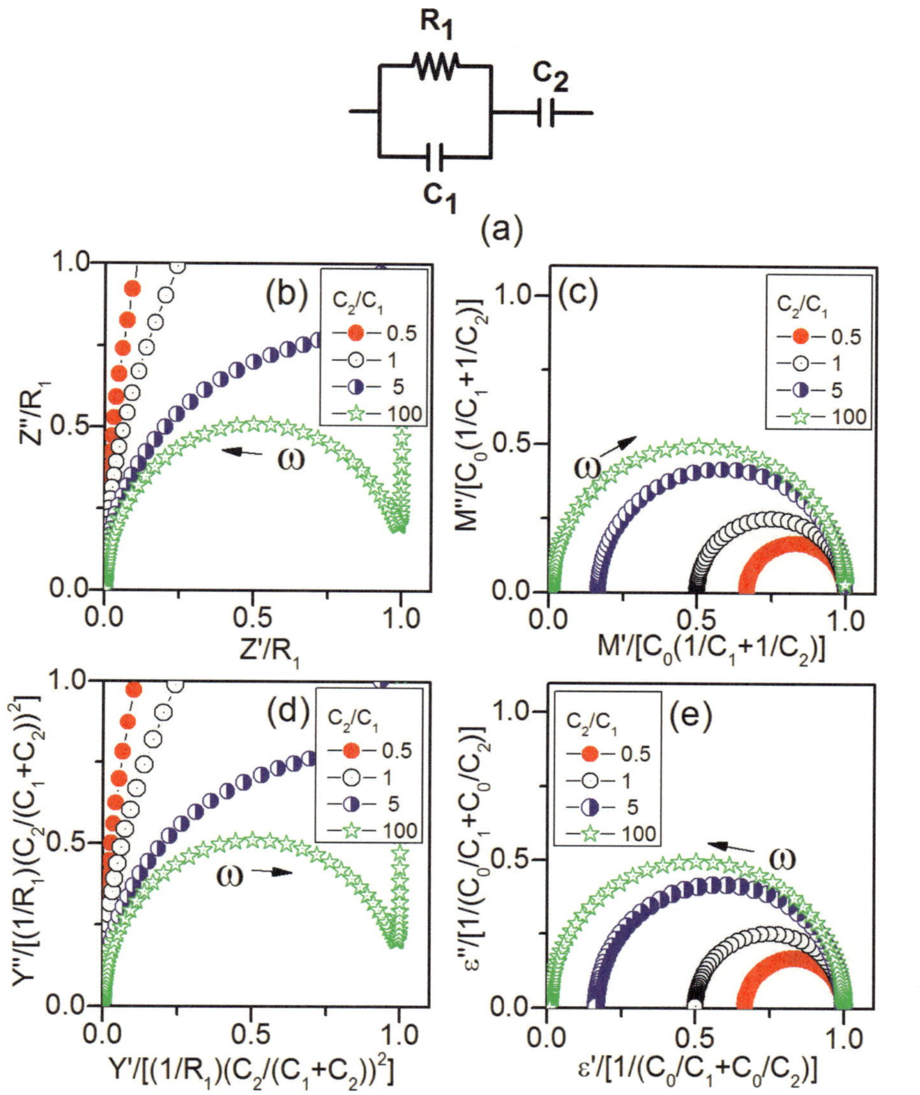

Figure 5.
(a) Equivalent circuit model containing series combination of parallel R_1C_1 and C_2. Normalized (b) Z''/R_1 vs Z'/R_1, (c) $M''/[C_0(1/C_1 + 1/C_2)]$ vs $M'/[C_0(1/C_1 + 1/C_2)]$, (d) $Y''/[(1/R_1)(C_2/(C_1 + C_2))^2]$ vs $Y'/[(1/R_1)(C_2/(C_1 + C_2))^2]$ and (e) $\varepsilon''/[1/(C_0/C_1 + C_0/C_2)]$ vs $\varepsilon'/[1/(C_0/C_1 + C_0/C_2)]$ for C_2/C_1 = 0.5, 1, 5 and 100.

3.4 Model 4: parallel combinations R_1C_1, R_2C_2 connected in series with CPE

Parallel combinations R_1C_1, R_2C_2 connected in series with a constant phase angle element (CPE) are shown in **Figure 6(a)**. The impedance of CPE is given as

$$Z_{CPE} = (\ Y_{CPE}\ )^{-1} = [\ A_0(j\omega)^{\Psi}]^{-1} = \frac{1}{A_0\omega^{\Psi}}\cos\left(\frac{\Psi\pi}{2}\right) - j\,\frac{1}{A_0\omega^{\Psi}}\sin\left(\frac{\Psi\pi}{2}\right) \tag{8}$$

Plot of imaginary part vs. real part of Z_{CPE} is a straight line with slope $\tan(\psi\pi/2)$, which remains constant as ω varies (hence the name CPE). For $\psi = 1$, the real part becomes zero and $Z_{CPE} = 1/(j\omega A_0)$ and the CPE behaves like an ideal capacitor having capacitance A_0. For $\psi = 0$, the imaginary part becomes zero and $Z_{CPE} = 1/A_0$

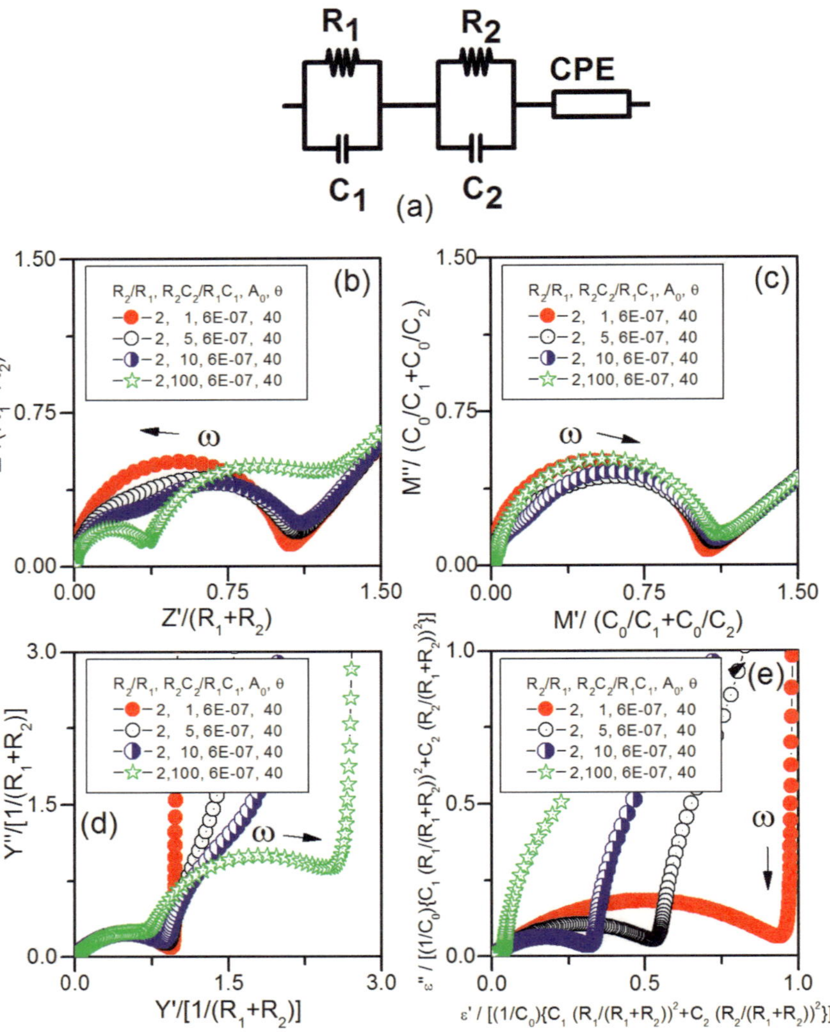

Figure 6.

(a) Equivalent circuit model containing series combination of parallel R_1C_1, parallel R_2C_2 and CPE. Normalized (b) $Z''/(R_1 + R_2)$ vs $Z'/(R_1 + R_2)$, (c) $M''/(C_o/C_1 + C_o/C_2$ vs $M'/(C_o/C_1 + C_o/C_2$, (d) $Y''/[1/(R_1 + R_2)]$ vs $Y'[1/(R_1 + R_2)]$ and (e) $\varepsilon''/[(1/C_o)\{C_1(R_1/(R_1 + R_2))^2 + C_2(R_2/(R_1 + R_2))^2\}]$ vs $\varepsilon'/[(1/C_o)\{C_1(R_1/(R_1 + R_2))^2 + C_2(R_2/(R_1 + R_2))^2\}]$ for $R_2/R_1 = 2$, $R_2C_2/R_1C_1 = 1, 5, 10, 100$, $A_o = 6 \times 10^{-7}$ and $\theta = 40$.

and the CPE behaves like an ideal register of value $1/A_0$ [6]. The values of Z′ and Z″ for the model shown in **Figure 6(a)** are given as [9, 21].

$$Z' = \frac{R_1}{1 + (\omega C_1 R_1)^2} + \frac{R_2}{1 + (\omega C_2 R_2)^2} + \left(\frac{1}{A_0 \omega^{\psi}}\right) \cos\left(\frac{\psi\pi}{2}\right) \tag{9}$$

$$Z'' = \frac{\omega C_1 R_1 R_1}{1 + (\omega C_1 R_1)^2} + \frac{\omega C_2 R_2 R_2}{1 + (\omega C_2 R_2)^2} + \left(\frac{1}{A_0 \omega^{\psi}}\right) \sin\left(\frac{\psi\pi}{2}\right) \tag{10}$$

A model comprising series combination of parallel R_1-CPE_1 and parallel R_2-CPE_2 is one of the models very widely used to represent the behaviour of a ceramic when the impedance plots have two depressed arcs. The reader is referred to [9] where this has been discussed by the authors in detail.

4. Experimental setup, measurements and results

Development of equivalent circuit model for representing the ceramic system $BaFe_xTi_{1-x}O_3$ (x = 0.05), prepared in our laboratory, is now described as an illustration. Sample was prepared by solid state synthesis method by taking $BaCO_3$ (Merck 99.5%), Fe_2O_3 (Merck 99.5%) and TiO_2 (Merck 99.5%) in appropriate amounts, mixing in acetone medium for 6 hours and calcining at 1100°C for 6 hours. The calcined powder was mixed with small amount of PVA binder and pressed into disc-like (dia 12 mm, thickness 1.5 mm) pellets using uniaxial hydraulic press with 60 kN pressure. These pellets were sintered in an electrical furnace (Lenton, made in Germany) where, first, binder was removed by raising the temperature to 500°C and holding for 2 hours and then increasing the temperature to 1250°C at 5°C/min and maintaining there for 10 hours followed by cooling it to room temperature. X-ray diffraction analysis revealed that the sample contained tetragonal and hexagonal phases in equal amounts. Impedance measurements were carried out as function of frequency (20 Hz to 1 MHz) at temperatures from 300 to 650 K. For this, the pellets were polished using emery papers of grade 1/0 and 2/0 and electroded on both sides using silver paste and cured at 600°C for 15 minutes.

The sample holder used to house the sample is schematically shown in **Figure 7(a)**. The spectroscopic and complex plane plots for M′ and M″ are shown in **Figure 7(b,c)**. The corresponding Z plots are not shown for brevity. The way an equivalent circuit model representing the data was developed is now described. A quick look at the M″ vs. M′ plot shown in **Figure 7(b)** reveals that there is no shift in the graph as well as no steeply rising high-frequency branch. Similar behaviour was seen in the Z plots also. Therefore, following the tips presented in Section 2, presence of series resistance or capacitance is ruled out. The plot is not a clear semicircular arc but is slightly depressed indicating the presence of more than one charge transfer processes in the system. As the sample contains two phases, we have a system comprising two types of grains, grain boundary and contact electrode interface. Therefore, an equivalent circuit model comprising four parallel RC's connected in series, where two RC's say R_1C_1 and R_2C_2 represent the two phases, R_3C_3 represents the grain boundary and R_4C_4 correspond to the sample-electrode interface, seems to be a plausible model. If we assume that $R_1C_1 < R_2C_2 < R_3C_3 < R_4C_4$ then, since electrode responses appear at low frequencies [1], R_1C_1 and R_2C_2 may be assigned to grains, R_3C_3 to grain boundary and R_4C_4 to electrode interface. The individual contributions from these RC's may be depicted by drawing tentative arcs (as shown by dotted lines in **Figure 7(b)**) making intercepts on the M′ axis at C_0/C_4, $C_0/C_4 + C_0/C_3$, $C_0/C_4 + C_0/C_3 + C_0/C_2$ and $C_0/C_4 + C_0/C_3 + C_0/C_2 + C_0/C_1$. As the plot is very symmetric, the intercepts and hence C's may be taken to be equal. By noting the values of intercepts from the graph, taking the value of C_0 as 0.6832×10^{-12}, noting the frequencies where the tentative arcs would peak and using the relation $\omega RC = 1$, the values of R' and C's are estimated as $R_1 = 362\ \Omega$, $R_2 = 1087\ \Omega$, $R_3 = 32{,}613\ \Omega$, $R_4 = 9944\ \Omega$ and $C_1 = C_2 = C_3 = C_4 = 4.88 \times 10^{-10}$ F. By using these as initial guesses, the values of the components were obtained accurately by running the CNLS program IMPSPEC.BAS developed by one of the authors [22] and being regularly used by us. These values were $R_1 = (0.77 \pm 0.01)$ kΩ, $C_1 = (0.38 \pm 0.01)$ nF, $R_2 = (3.43 \pm 0.02)$ kΩ, $C_2 = (0.44 \pm 0.01)$ nF, $R_3 = (10.87 \pm 0.09)$ kΩ, $C_3 = (1.08 \pm 0.06)$ nF, $R_4 = (15.36 \pm 0.26)$ kΩ and $C_4 = (5.81 \pm 0.09)$ nF. The M′ and M″ values corresponding to the fitted RC's are also shown in **Figure 7(c)**.

Use of different models for impedance spectroscopy of few other ceramics is briefly described now. Analysis of data for $SrTiO_3$ borosilicate glass ceramics having

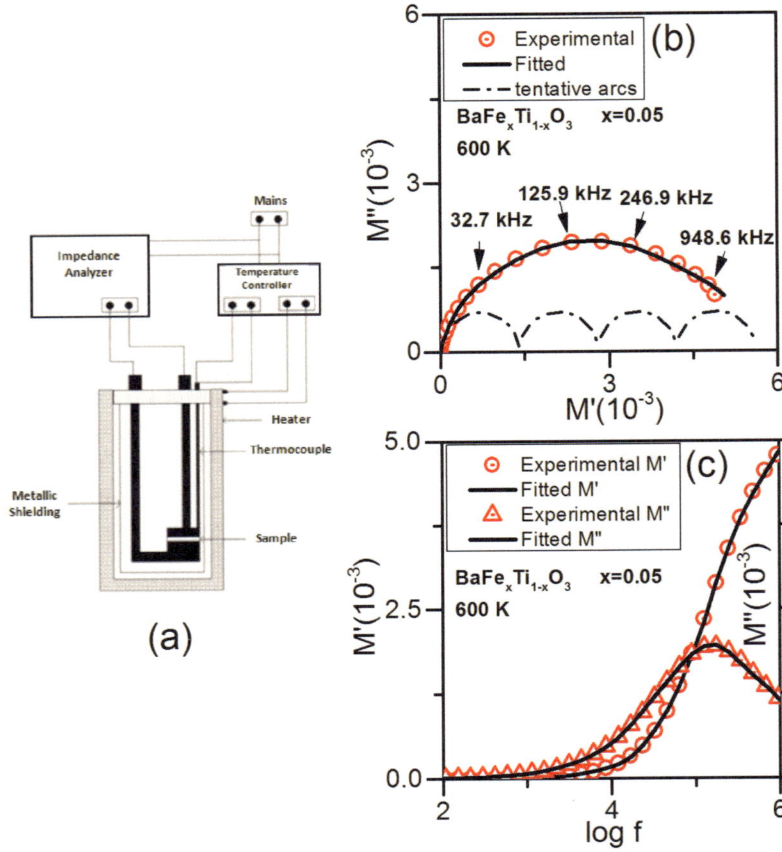

Figure 7.
(a) Experimental Setup for Impedance Measurements. Experimental and fitted values of (b) M″ vs M′ for $BaFe_xTi_{1-x}O_3$ (x = 0.05) at 600 K and (c) M′, M″ vs log f for $BaFe_xTi_{1-x}O_3$ x = 0.05 at 600 K.

steeply rising low-frequency branch in Z″ vs. Z′ plots has been done by Pandey et al. [17]. $Ba_{1-x}La_xTi_{1-x}Co_xO_3$ ($x \leq 0.20$) ceramic showed a shifted arc in Z" vs. Z' plot and was analysed by using equivalent circuit model involving a series resistance [19]. In this system, presence of two PTCR components, one due to grain boundary as usual and the other in grains was observed by impedance spectroscopy. Study of semiconducting $BaTiO_3$ was carried out by using a model containing series resistance [18]. Ceramic system $Ca_{1-x}Y_xTi_{1-x}Co_xO_3$ could be represented by an equivalent circuit model containing two parallel RC's and a CPE connected in series [21]. Models useful for representing magnetic ceramic are not covered in this paper. A detailed analysis of models involving inductive component L is available in [11] where magnetic ceramic YIG has been studied by using impedance spectroscopy.

While using the impedance spectroscopy, it is important to keep in mind that the observed impedance values are those what the impedance analyser sees at its input points. These values include the contributions from the connecting leads, cables, sample holder and electrodes. It has been found that the electrode-sample contact behaviour changes with the nature of the sample [15]. The sample-electrode contact effects have been studied in detail in [23–35]. A method has been proposed for removing the sample holder contributions [23], which involves doing the complete impedance spectroscopic analysis without the sample, finding an empirical functional relation for the impedance behaviour and then subtracting this from the results obtained with sample placed in the sample holder. Presently available analysers provide with a lead correction step before an actual measurement is done.

For separating out contact contributions, impedance spectroscopic studies by repeating the measurements with changed electrodes have been reported [36]. Measurements by using samples of different thicknesses might also be useful. Various sources of errors in impedance measurements have been discussed in [6].

5. Conclusion

Some equivalent circuit models most useful for impedance spectroscopic studies of electronic ceramics and their simulated immittance behaviour are discussed. In order to facilitate prompt development of equivalent circuit models, few extremely helpful steps have been summarised. A comparison of the experimental plots with simulated ones provides a clue for inclusion of certain lumped components in the model, e.g. a right shift in the Z″ vs. Z′ plot indicates presence of series resistance and a shift in M″ vs. M′ plot indicates presence of series capacitance in the model. The models are not unique. The most appropriate model may be arrived at by looking at the immittance plots in more than one formalisms (such as Z and M) for all the experimental parameters such as composition, temperature etc, taken together, and considering the possibilities of processes present/dormant, emerging or dominating in the system as some variables such as temperature are altered. The procedure of impedance spectroscopic modelling is illustrated by analysing the impedance data on the ceramic system $BaFe_xTi_{1-x}O_3$ (x = 0.05) containing two phases.

Acknowledgements

The financial support received by one of the authors (SP) from IIT(BHU) in the form of Teaching Assistantship is gratefully acknowledged.

Author details

Shukdev Pandey[1*], Devendra Kumar[1], Om Parkash[1] and Lakshman Pandey[2]

1 Department of Ceramic Engineering, IIT(BHU), Varanasi, India

2 Department of Physics and Electronics, Rani Durgavati University, Jabalpur, India

*Address all correspondence to: spandey.rs.cer12@itbhu.ac.in

References

[1] Moulson AJ, Herbert JM. Electroceramics. 2nd ed. England: John Wiley & Sons Ltd.; 2003

[2] Buchanan RC. Ceramic Materials for Electronics. 3rd ed. New York: CRC Press; 2004

[3] Sebastian MT. Dielectric Materials for Wireless Communications. 1st ed. Oxford: Elsevier Ltd; 2008

[4] Callister WD Jr. Materials Science and Engineering : An Introduction. 7th ed. New York: John Wiley & Sons Inc; 2007

[5] MacMillan PW. Glass Ceramics. 2nd ed. London: Academic Press; 1979

[6] Barsoukov E, Macdonald JR. Impedance Spectroscopy Theory, Experiment, and Applications. 2nd ed. New Jersey: John Wiley & Sons; 2005

[7] Jonscher AK. Dielectric Relaxation in Solids. 1st ed. London: Chelsea Dielectric Press; 1983

[8] Pandey L, Parkash O, Katare RK, Kumar D. Equivalent circuit models for electronic ceramics. Bulletin of Materials Science. 1995;**18**:563-576. DOI: 10.1007/BF02744842

[9] Pandey S, Kumar D, Parkash O, Pandey L. Equivalent circuit models using CPE for impedance spectroscopy of electronic ceramics. Integrated Ferroelectrics. 2017;**183**:141-162. DOI: 10.1080/10584587.2017.1376984

[10] Sinclair DC, West AR. Effect of atmosphere on the PTCR properties of $BaTiO_3$ ceramics. Journal of Materials Science. 1994;**29**:6061-6068. DOI: 10.1007/BF00354542

[11] Katare RK, Pandey L, Dwivedi RK, Parkash O, Kumar D. A novel approach based on impedance spectroscopy for measurement of magnetic permeability of ceramics. Indian Journal of Engineering & Materials Sciences. 1999; **6**:34-42. http://nopr.niscair.res.in/handle/123456789/21976

[12] Katare RK. Application of Impedance Spectroscopy in the Study of Electronic Ceramics [thesis]. Jabalpur, India: Rani Durgavati University; 1997

[13] Chaitanya P. Impedance Spectroscopy and RF Pulse Response of Piezoelectric Materials. [Thesis]. Jabalpur, India: Rani Durgavati University; 2009

[14] Chaitanya P, Shukla A, Pandey L. Determination of equivalent circuit model components of piezoelectric materials by using impedance spectroscopy. Integrated Ferroelectrics. 2014;**150**:88-95. DOI: 10.1080/10584587.2014.874274

[15] Chaitanya P, Mishra R, Ahirwal P, Shukla A, Thakur OP, Pandey L, et al. Study of temperature dependence of electrode–glass ceramic interface using impedance spectroscopy. Integrated Ferroelectrics. 2015;**159**:121-126. DOI: 10.1080/10584587.2015.1033048

[16] Chaitanya P, Thakur OP, Kumar V, Shukla A, Pandey L. Equivalent circuit model of a $PbZr_{0.6}Ti_{0.4}O_3$ ceramic using impedance spectroscopy. Journal of Ceramic Processing Research. 2011;**12**: 247-258

[17] Pandey L, Katare RK, Parkash O, Kumar D, Thakur OP. Complex impedance analysis of electronic ceramics showing steeply rising impedance pattern at low frequency. Indian Journal of Pure & Applied Physics. 1998;**36**:228-235

[18] Maiti HS, Basu RN. Complex plane impedance analysis for semiconducting barium titanate. Materials Research

Bulletin. 1986;**21**:1107-1114. DOI: 10.1016/0025-5408(86)90227-8

[19] Pandey L, Katare RK, Parkash O, Kumar D. Evidence of two ferroelectric PTCR components in valence compensated ceramic system $Ba_{1-x}La_xTi_{1-x}Co_xO_3$. Bulletin of Materials Science. 1997;**20**:933-947. https://www.ias.ac.in/article/fulltext/boms/020/07/0933-0947

[20] Kumar V, Thakur OP, Pandey L, Guimaraes AP, Goel A, Parkash O, et al. Modeling of electrical behavior of $La_{0.7}Ca_{0.3}MnO_3$ ceramic using impedance spectroscopy. Modern Physics Letters B. 2005;**19**:697-706. DOI: 10.1142/S0217984905008621

[21] Katare RK, Pandey L, Thakur OP, Parkash O, Kumar D. Equivalent circuit model of $Ca_{1-x}Y_xTi_{1-x}Co_xO_3$ using impedance spectroscopy. Modern Physics Letters B. 2003;**17**:339-346. DOI: 10.1142/S0217984903005238

[22] Pandey L. In: Workshop on Use of Computers in Teaching Physics; Partly Sponsored by ICTP; 3–9 December 1992; Jabalpur, India

[23] Pandey L, Parkash O, Kumar D. On sample holder corrections in the complex impedance analysis of electronic ceramics. Indian Journal of Pure & Applied Physics. 1996;**34**:28-33

[24] Edwards DD, Hwang JH, Ford SJ, Mason TO. Experimental limitations in impedance spectroscopy: Part V. Apparatus contributions and corrections. Solid State Ionics. 1997;**99**:85-93. DOI: 10.1016/S0167-2738(97)00206-3

[25] Basu RN, Maiti HS. PTC behaviour of semiconducting BaTiO3 ceramics. Transactions of the Indian Ceramic Society. 1986;**45**:140-146. DOI: 10.1080/0371750X.1986.10822812

[26] Deguin A, Moretti P, Boyeaux JP. Electrical properties of Nb doped $BaTiO_3$ crystal with special emphasis on contact influence. Ferroelectrics. 1980;**26**:761-764. DOI: 10.1080/00150198008008166

[27] Serghei A, Tress M, Sangoro JR, Kremer F. Electrode polarization and charge transport at solid interfaces. Physical Review B. 2009;**80**:184301-184305. DOI: 10.1103/PhysRevB.80.184301

[28] Sato H, Manghna MH, Lienert BR, Weiner A. Effects of sample-electrode interface polarization on the electrical properties of partially molten rock. Journal of Geophysical Research. 1986;**91**:9325-9332. DOI: 10.1029/JB091iB09p09325

[29] Pizzitutti F, Bruni F. Electrode and interfacial polarization in broadband dielectric spectroscopy measurements. Review of Scientific Instruments. 2001;**72**:2502-2504. DOI: 10.1063/1.1364663

[30] Ahmed R, Reifsnider K. Study of influence of electrode geometry on impedance spectroscopy. International Journal of Electrochemical Science. 2011;**6**:1159-1174

[31] Costa MEV, Mantas PQ, Baptista JL. Effect of electrode alterations on the a.c. behaviour of Li_2O–ZnO humidity sensors. Sensors and Actuators B. 1995;**26-27**:312-314. DOI: 10.1016/0925-4005(94)01608-K

[32] Hsieh G, Ford SJ, Mason TO, Pederson LR. Experimental limitations in impedance spectroscopy: Part I—Simulation of reference electrode artifacts in three-point measurements. Solid State Ionics. 1996;**91**:191-201. DOI: 10.1016/S0167-2738(96)83019-0

[33] Hsieh G, Mason TO, Pederson LR. Experimental limitations in impedance spectroscopy: Part II–Electrode artifacts in three-point measurements on Pt/YSZ.

Solid State Ionics. 1996;**91**:203-212. DOI: 10.1016/S0167-2738(96)83020-7

[34] Hsieh G, Mason TO, Garboczi EJ, Pederson LR. Experimental limitations in impedance spectroscopy: Part III. Effect of reference electrode geometry position. Solid State Ionics. 1997;**96**: 153-172. DOI: 10.1016/S0167-2738(97) 00073-8

[35] Hwang JH, Kirkpatrick KS, Mason TO, Garboczi EJ. Experimental limitations in impedance spectroscopy: Part IV. Electrode contact effects. Solid State Ionics. 1997;**98**:93-104. DOI: 10.1016/S0167-2738(97)00075-1

[36] Maso N, Beltran H, Cordoncillo E, Escribano P, West AR. Electrical properties of Fe-doped $BaTiO_3$. Journal of Materials Chemistry. 2006;**16**: 1626-1633. DOI: 10.1039/B515834F